T0215191

Digital Sampling

Digital Sampling is the first book about the design and use of sampling technologies that have shaped the sounds of popular music since the 1980s.

Written in two parts, *Digital Sampling* begins with an exploration of the Fairlight CMI and how artists like Kate Bush and Peter Gabriel used it to sample the sounds of everyday life. It also focuses on E-mu Systems and the use of its keyboards and drum machines in hip-hop. The second part follows users across a range of musical worlds, including US/UK garage, indie folk music, and electronic music made from the sounds of sewers, war zones, and crematoriums.

Using material from interviews and concepts from the field of Science and Technology Studies (STS), *Digital Sampling* provides a new and alternative approach to the study of sampling and is crucial reading for undergraduates, postgraduates, and researchers from a wide range of disciplines, including music technology, media, communication, and cultural studies.

Paul Harkins is a lecturer in the Music Department at Edinburgh Napier University.

Digital Sampling
The Design and Use of Music Technologies

Paul Harkins

Routledge
Taylor & Francis Group

NEW YORK AND LONDON

First published 2020
by Routledge
52 Vanderbilt Avenue, New York, NY 10017

and by Routledge
2 Park Square, Milton Park, Abingdon, Oxon, OX14 4RN

Routledge is an imprint of the Taylor & Francis Group, an informa business

Library of Congress Cataloging-in-Publication Data
Names: Harkins, Paul (Paul Michael) author.
Title: Digital sampling : the design and use of music technologies / Paul Harkins.
Description: New York, NY : Routledge, 2020. | Includes bibliographical references and index.
Identifiers: LCCN 2019016720 (print) | LCCN 2019018021 (ebook) | ISBN 9781351209960 (master ebk) | ISBN 9781351209953 (pdf) | ISBN 9781351209939 (mobi) | ISBN 9781351209946 (epub3) | ISBN 9781138577510 (hbk : alk. paper) | ISBN 9780815381648 (pbk : alk. paper)
Subjects: LCSH: Sampler (Musical instrument)—History. | Electronic musical instruments—History. | Synthesizer (Musical instrument)—History.
Classification: LCC ML1092 (ebook) | LCC ML1092 .H3 2020 (print) | DDC 786.7—dc23
LC record available at https://lccn.loc.gov/2019016720

ISBN: 978-1-138-57751-0 (hbk)
ISBN: 978-0-8153-8164-8 (pbk)
ISBN: 978-1-351-20996-0 (ebk)

Typeset in Sabon
by Swales & Willis Ltd, Exeter, Devon, UK

Contents

Figures

Prelude/Acknowledgements

This is a book about sampling in popular music but it is really about technologies and what people do with them. It will not give you a definitive answer about what the first sampler was. Instead, it will show that instruments now considered the first samplers were designed to do very different things. Search on Google or Wikipedia and you'll find that sampling in music is about taking sounds from pre-existing recordings. I want to tell a different story about sampling that shows how the practices associated with it have changed and how sampling is now, and perhaps always has been, about the use of any pre-recorded, digitally reproduced sound. It is a story that will challenge myths: 'The Mellotron was an early sampler'; 'The Fairlight was the first sampler'; 'The Emulator was the first affordable sampler'. It is a story that I have written primarily for students and researchers, but it should hopefully be of interest to anyone who wants to understand how popular music has changed over the last four decades.

I started doing the interviews that form the basis of this book in 2008 and there are a lot of people to thank for helping me in different ways as I transformed it from a PhD thesis to a book. Colleagues at Edinburgh Napier University have been supportive as we juggle the growing demands of research, teaching, and administrative work: Craig Ainslie, Nicholas Ashton, Chris Atton, Katrina Burton, Ken Dempster, Willie Duff, Paul Ferguson, Andrew Frayn, John Hails, Dave Hook, Michael Harris, Pauline Miller Judd, Linda Leyden, Rune Lilledal Hansen, Haftor Medbøe, Zack Moir, Anne Schwan, Renée Stefanie, and Bryden Stillie. Simon Frith ran a popular music seminar at the University of Edinburgh for a number of years that helped me develop ideas and a global network of academic friends. Thank you Simon and hats off to Melissa Avdeef, Adam Behr, Matt Brennan, Evangelos Chrysagis, Kieran Curran, Kyle Devine, Ninian Dunnett, Mary Fogarty Woehrel,

Tami Gadir, Mark Percival, Luis Sanchez, Arnar Eggert Thoroddsen, Emma Webster, Tom Western, Sean Williams, and Richard Worth. I have presented the ideas in this book at seminars, workshops, and conferences and want to thank Annie Jamieson and James Mooney for kickstarting the OESM network at the National Science and Media Museum in Bradford, where I got close(r) to a Fairlight CMI. Francois Ribac invited me to Dijon and Paris where I was lucky enough to hang out with a great bunch of scholarly people at IRCAM including George Brock-Nannestad, Jens Gerrit Papenburg, Jacob Smith, and Basile Zimmerman. I would also like to thank colleagues/friends/acquaintances who are members of IASPM, ASARP, and SPAN, including Martin Cloonan and John Williamson. I am grateful to Kyle Devine, James Sumner, Tom Western, and Justin Williams for taking the time to look over drafts of specific chapters and offering useful comments. Any mistakes and misjudgements are mine and mine alone.

I would like to thank the designers and distributors of digital synthesizer and sampling instruments who answered questions that helped with my research: Dan Coren and Harry Mendell of Computer Music Incorporated; Cameron Jones of New England Digital; Roger Linn; Dany, and Peter Dean at Publison in Paris; Marco Alpert of E-mu Systems; Peter Vogel, Peter Farleigh, and Peter Wielk at Fairlight Instruments in Sydney; Jonathan Cole and Michael Kelly of Syco Systems; and Steve Rance at Fairlight US. Thank you also to the technology users (and non-users) I interviewed: Kenny Anderson, Richard Burgess, Jay Burnett, Ziggy Campbell, Ian Curnow, Bill Drummond, Todd Edwards, Phil Harding, Matthew Herbert, Trevor Horn, J.J. Jeczalik, Keith Le Blanc, Marc Leclair, Paul D. Miller, Tommy Perman, Oliver Sabin, Kevin Sim, Carl Stone, and Drew Wright. I'd also like to thank everyone who helped with my archival research including library staff at Edinburgh Napier and Edinburgh University as well as John Twyman in Sydney and R. of retrosynthads.blogspot.com.

To old friends in Edinburgh, and now East Lothian, thanks for keeping in touch when the writing of this book decreased my alcohol intake. As we get older (and wiser?) the struggle for fun continues: Rod Aitchison, David Cross, Iain Harron, Grant McClory, Robert Scott, Adam Taylor, and Andrew Veitch. To the ones who were there on Saturday 21 May 2016: GGTTH. To new friends in Fife: Scott, David, Alan, and Jackie at Positive Change Yoga; Kirsty and Anthony at Kangus Coffee House; all the Beveridge Parkrunners and Kirkcaldy Wizards: Gary, John, Malcolm, Neil, Jock, Rebecca, Michael, Ian, Michelle, Heather, Kevin, Lindsay, Erin, Ronnie, and Paul for showing me the strength of weak ties and reminding me that life is a marathon, not a sprint. Finally, and most importantly, for their love and support: Mum, Kate, Stuart, Hannah, Emma, Craig, Finn, Eoin, Willie, Leila, Paul, Simon, Angelique, and all my family. Finally, finally to Allison and Paul Jr: I love you both.

Thanks to all the staff at Routledge who have helped me turn this manuscript into the book you have in front of you: Lara Zoble for asking if I wanted to write a book in the first place; Claire Margerison, Hannah Rowe, and Shannon Neill for their editorial guidance; and Tamsin Ballard and Gayle Green for the copyediting and help with the final stages. Early versions of some material in the book first appeared in the following articles: 'Transmission Loss and Found: The Sampler as Compositional Tool', *Journal on the Art of Record Production*, April 2009; 'Appropriation, Additive Approaches and Accidents: A Case Study of the Sampler as Compositional Tool and Recording Dislocation', *IASPM@Journal*, 1:2, 2010; *Microsampling: From Akufen's Microhouse to Todd Edwards and the Sound of UK Garage in Musical Rhythm in the Age of Digital Reproduction*, ed. Anne Danielsen (Farnham: Ashgate, 2010); 'Following the Instruments, Designers, and Users: The Case of the Fairlight CMI', *Journal on the Art of Record Production*, October 2015.

Introduction

In January 2011, the National Association of Music Merchants (NAMM) organised a trade show in California, where a prototype of a musical instrument was launched. The show attracted more than 1,500 companies including Akai, Fender, Roland, and Yamaha, who exhibited their products. At booth 1252 in Hall E, representatives of an Australian company called Fairlight Instruments demonstrated the Fairlight CMI-30A. The promotional materials described it as: 'a unique instrument, combining the latest technology with the look and feel of the original Fairlight CMI. It achieves the classic Fairlight sound that defined the music of the eighties' (Fairlight 2011a). The CMI-30A was not a new musical instrument but the thirtieth anniversary edition of an older one, the Fairlight Computer Musical Instrument (CMI). Often described as the first digital sampler, the Fairlight CMI was primarily a digital synthesizer and computer workstation. Launched in 1979, it was one of the technologies that were used to re-shape the practices of music making and the sounds of popular music in the 1980s. The Fairlight CMI could be found in university departments, professional recording studios, and institutions like the British Broadcasting Corporation (BBC). It was expensive and those musicians who could afford one were part of an elite group that included Kate Bush, Peter Gabriel, Herbie Hancock, and Stevie Wonder. There were other key users who were not owners. Afrika Bambaataa, Arthur Baker, and John Robie, for example, gained late-night access to a New York recording studio equipped with a Fairlight and used one of its pre-recorded orchestral sounds on 'Planet Rock' (1982).

Thirty years after the introduction of the Fairlight CMI, Fairlight Instruments were quick to claim credit for the ways in which their instrument transformed the technological processes of making music.[1] The advertising brochure for the CMI-30A stated:

> When the Fairlight CMI arrived on the scene in the eighties it changed the way we make music, forever. Today every sampler, digital synthesizer, and audio

1

> workstation can trace its lineage back to this legendary machine. Known for its solid, hand-built quality, and iconic sounds, the Fairlight CMI holds a special place in history and the hearts of musicians everywhere.
>
> Fairlight (2011a)

Users had to learn new approaches to making music with a personal computer during early experiments with the Fairlight CMI. The pre-recorded library sounds were stored on 8-inch floppy disks. A light-pen was used to choose options from an on-screen menu. Typing a command on the QWERTY keyboard was as important as playing a note on the piano keyboards. In a review of the Fairlight CMI in 1981, journalist Richard Elen outlined some of the issues involved in learning a programming language to communicate with a computer: '[H]alf the battle with any computer system would seem to be getting to grips with the 'man-machine interface' – talking to the bloody thing without it informing you of a **COMMAND SYNTAX ERROR**!' (p. 45). Instruction manuals were included and the difficulties of translating technical language resulted in ad-hoc tutorials over the telephone with its designers. When the Fairlight CMI arrived on the scene in the 1980s, users were not always sure what to do with this technology.[2]

Interpretative Flexibility: The Social Construction of Sampling Technologies

> Of course, a sampler is a musical instrument as well as a production tool. It allows you control over any sound. You can make music out of a toilet and a [Gheorghe] Zamfir record with a sampler! I think samplers have been considered musical instruments for at least the past 25–30 years or so haven't they?
>
> (Aaron Funk aka Venetian Snares, quoted at M3 Event 2012).

One of the problems for anyone studying the historical and contemporary uses of sampling technologies is that the field of organology – the academic study of musical instruments – has been dominated by the instruments of western art music and those designed before the introduction of electricity, with little focus on the contexts of their use.[3] To understand the relationship between the inventors of music technologies and their users, researchers might be advised to enter the field of Science and Technology Studies (STS). Here, sociologists have challenged both technological determinism and the emphasis on inventors as heroic geniuses that was prevalent in the histories of technology written in the twentieth century.[4] In STS, you can choose from a rich mix of concepts and engage with the work of scholars like Wiebe Bijker and Trevor Pinch who are identified with the social construction of technology (SCOT) approach to understanding the design and use of technologies. These range from bicycles and Bakelite (Bijker, Hughes, and Pinch 1987) to cars and contraceptives (Oudshoorn and Pinch 2003). Other STS scholars such as Madeleine Akrich and Bruno Latour are more closely associated with Actor Network Theory (ANT) than the SCOT approach to studying technologies. ANT ascribes agency to both humans and non-humans, a flat ontology rejected by social constructivists like Bijker and Pinch who write: '[T]he typical ANT step of making no ontological distinction between human and non-human actants is not made in SCOT' (Bijker and Pinch 2012, p. xxii). What all these scholars share is a 'focus on what social groups and actor networks actually say and do with technology' (p. xxi). My book continues the work of SCOT by focusing on what the designers of sampling technologies were trying to do in the 1970s and 1980s. I also place a strong emphasis on what *users* were doing (and not doing) with these technologies as they became embedded in the everyday practices of music making.

One of the key concepts from SCOT that is helpful for understanding the ways music technologies such as the Fairlight CMI were used in a variety of different contexts and in ways unforeseen by their designers is that of *interpretative flexibility* (Pinch and Bijker 1984). This is the idea that new technologies are designed and developed, and undergo changes as a result of use (or non-use) by 'relevant social groups' before arriving at a period of stability and closure where a dominant form of the technology emerges.[5] For example, in the 1960s, engineers Don Buchla and Robert Moog developed analogue synthesizers but with different designs. Moog's modular synthesizers proved to be more successful because the piano keyboard was used as its interface and was already familiar to groups of users like rock musicians. This is one of the reasons why the keyboard-based synthesizer became the dominant form of the technology in the 1970s and 1980s.[6] One problem, though, with the concept of 'interpretative flexibility' is its emphasis on stabilisation. The design and use of music technologies might not end with any form of closure. For instance, interest in modular rack-based synthesizers has been growing in recent years, as some musicians grow frustrated with computers and software.[7] Where Pinch and Bijker focus on closure in the design of technological artefacts, this book is about the opposite: the many types of sampling technologies – digital synthesizers, sampling keyboards, sampling drum machines, rack-based samplers, PCs and laptops with software samplers – that have been used, and continue to be used, for making music.

Defining Sampling

Since the mid-1980s, sampling and samplers have been closely associated with the genres of hip-hop and electronic dance music (EDM), and become synonymous with a particular musical practice: the re-use of an excerpt from a pre-existing recording in a new recording.[8] Sampling has often been defined in a reductive way as a form of musical quotation or, to its detractors, theft.[9] In contrast to this narrow definition of sampling, I explore a diverse range of social, musical, and technological practices that have shaped both historical and contemporary approaches to sampling. These include the use of sampling technologies to imitate the sounds of acoustic instruments and, what advertisements in the early 1980s described as, the sampling of 'external' or 'natural' sounds. As sampling became identified primarily with appropriation, a concept that previously referred to the digitalisation of sound is now used ahistorically to describe any re-use of pre-existing sound: The Beatles are said to have used 'samples' on 'Being for the Benefit of Mr. Kite!' (1967) when George Martin and his engineer dubbed recordings of Victorian steam organs onto magnetic tape and spliced short excerpts together. Sugarhill Gang are said to have 'sampled' Chic's 'Good Times' even though no digital samplers were used in the making of 'Rapper's Delight' (1979). The use of analogue technologies like tape in the 1960s and the re-playing of sounds using acoustic and electric instruments in the 1970s needs to be disentangled from the discourses around the use of digital technologies that developed in the 1980s. The definition of sampling that runs through this book is the use of digital technologies to record, store, and reproduce *any* sound. The widespread adoption of digital technologies in the production of music since the 1980s has led Eliot Bates to explain how: 'In contemporary computer-based audio recording, every moment of recorded sound is essentially a 'sample.' Thus, rock 'n' roll, country, blues, and classical genres not traditionally associated with sampling – are now sample-based musics' (2004, p. 283). This refers specifically to the process of digitising sound. What I do in this book is follow the users of sampling technologies across a range of genres including pop, hip-hop, folk, microhouse, and UK garage to understand the multiplicity of approaches to using digital samples that have developed over the last four decades.

While a sample is often thought of as an excerpt or something representative of a larger entity, the word *sampler* comes from the Latin *exemplar*, which means a model or pattern. From the seventeenth century onwards, presentation samplers were created and used as a reference point for those learning the art of embroidery to copy stitches and patterns. The Philadelphia Museum of Art has one of the largest collections of presentation samplers in the world. In 1912, a confectionary company based in the city, Whitman's, began producing a selection of its best-selling chocolates as the Whitman's Sampler with an embroidery pattern on the box. In the early 1980s, the designers at E-mu Systems, partly inspired by this product, came up with Sampler as the working title for its sampling keyboard, the Emulator (1981). This was because it could re/produce a variety of sounds. Even though sample times increased significantly during the 1980s and software samplers can now record hours of data, a sample is often thought of as a short excerpt of sound. As well as the use of the word in marketing to denote a specimen or a small example of a larger range of products, sampling has been used in both qualitative and quantitative methods of research since the development of a sampling theory approach to statistics in the eighteenth century. It has also been applied more recently to demographic analysis and political polling since the twentieth century. This idea of a representative sample, or the gathering of an incomplete collection of data, also relates to the principles of digital audio that underpin the design of sampling technologies.

As digital computing technologies were developed in the 1940s and companies like IBM (International Business Machines) introduced computers into the workflows of US corporations in the 1950s (Campbell-Kelly et. al. 2018), experiments in producing sounds with computers took place in research institutes, communications laboratories, broadcasting corporations, and university departments.[10] In his work as director of the Behavioral Research Laboratory at Bell Telephone Laboratories in New Jersey, Max Mathews used an IBM 7090 mainframe computer and the programming languages *Music I* to *Music V* (1957–1968) to make music. One of the problems he was trying to solve was: 'How can the numbers with which a computer deals be converted into sounds the ear can hear?' (Mathews 1963, p. 553). The answer was that binary digits needed to be converted into analogue signals using a digital-to-analogue converter (DAC). To convert sounds or analogue signals into a sequence of numbers, an analogue-to-digital converter (ADC) or a sampler needed to be used (Mathews 1969). Based on the sampling theorem developed by Harry Nyquist at Bell Laboratories in the 1920s and the later work of Claude E. Shannon in the 1940s, the waveform of an analogue sound was converted into a digital signal at a rate of 44,100 samples per second.[11] Sampling was, and is, the conversion of sound from an analogue signal into a digital one. However, the definition of sampling has expanded along with changing musical practices. As Thom Holmes explains: '[S]ampling can refer to the sampling rate of sounds that are directly synthesized by a computer or the digital reproduction of externally generated sounds' (2016, p. 309). The Fairlight CMI is now associated with the sampling of external sounds but the priority of its designers, Peter Vogel and Kim Ryrie, in the 1970s was the generation of digitally synthesized sounds to replicate the sounds of acoustic instruments. It was users who experimented with recording, storing, and reproducing sounds digitally: a musical and technological practice that became known as *sampling*.[12]

A Short History of Digital Synthesizers/An Early History of Sampling Instruments

The design and development of digital synthesizer technologies in the 1970s was made possible by the wider availability of minicomputers and increases in the power of microprocessor technology. It was also the indirect result of experiments in previous decades with analogue technologies – magnetic tape and modular synthesizers – in avant-garde musical worlds.[13] At the San Francisco Tape Music Center in the early 1960s, Ramon Sender, Morton Subotnick, and Don Buchla developed the Buchla 100 series Modular Electronic Music System, or 'Buchla Box', as a device for the mass production/consumption of electronic music. It also provided a solution to the problems of making incidental music in recording studios, with found sounds on magnetic tape; cutting and splicing pieces of tape together was time consuming. Voltage-controlled synthesizers could generate sounds electronically but were difficult to control without a keyboard. As the keyboard became the dominant interface in the 1970s, some users expressed disappointment that the potential of the modular synthesizer as a means of creating new sounds had not been fully explored. According to Bernie Krause, who began experimenting with a Buchla Box as a student at the San Francisco Tape Music Center, they were 'witnessing the evolution of the synthesizer from an instrument that could produce a variety of unknown sounds to one that reproduced a standard package of familiar sounds' (quoted in Pinch and Trocco 2002, p. 130). The 'interpretative flexibility' of synthesizer technologies was thought to have closed. As well as re-creating sounds without magnetic tape, the Buchla Box, like earlier synthesizers, had also been designed to imitate the sounds of acoustic instruments.[14] Those using analogue synthesizers to do so discovered that the sounds of musical instruments performed by humans were not easy to copy and instrument designers turned to digital synthesis as a way of trying to gain more control over the parameters of sound.[15]

In 1972, composer, Jon Appleton, and engineers, Sydney Alonso and Cameron Jones, began collaborating on the Dartmouth Digital Synthesizer in the Thayer School of Engineering at Dartmouth College. Inspired by the work of Bob Moog, Max Mathews, and John Chowning, the designers used a 16-bit microprocessor called ABLE to develop a performance instrument using digitally synthesized sounds. In 1977, Alonso and Jones formed New England Digital Corporation and, in June 1978, after a meeting with Mathews at Bell Laboratories and input from Appleton, they renamed the instrument the Synclavier. Fifteen were sold at a cost of $13,500US (Appleton 1989). The digital memory recorder it contained was an important feature, anticipating the development of MIDI[16] sequencing and the Digital Audio Workstation (DAW).[17] After witnessing the interest in the light-pen and user sampling technologies that were part of the Fairlight CMI, the designers of the Synclavier decided to give its users the opportunity to digitally record, store, and playback sounds of their own. Alonso explained:

> [Fairlight Instruments] would show up at AES [Audio Engineering Society] shows and they'd say, 'Well, let's draw a picture of a Volkswagen on the screen and then we'll play that wave,' and lo and behold, the public bought it, so all of a sudden the idea was that we want to do sampling – and this was a very strong market force, so we were forced to develop the sampling unit.
> (quoted in Chadabe 1997, p. 186)

Appleton left the company in 1979 to return to composition and education, and Berklee College of Music graduate, Bradley Naples, joined to develop a marketing strategy. New England Digital launched the Synclavier II in 1980 and added the option of monophonic sampling two years later.[18] Primarily a digital synthesizer, they designed the Synclavier to give users more control over the timbre of sounds than was possible with analogue synthesizers.[19] Their inclusion of a function to digitally record external sounds on the Synclavier II was a commercially driven decision to compete with the Fairlight CMI.

Along with the Fairlight CMI, other engineers were developing computer-based music technologies in the 1970s that enabled users to experiment with the digital reproduction of externally recorded sounds. In the Presser Electronic Music Studio at the University of Pennsylvania, Dan Coren and Harry Mendell designed an instrument that became known as the Computer Music Melodian. In 1975, they formed a company called Computer Music Incorporated to market the product.[20] The hardware and software allowed users to digitally record their own sounds and play them on the keyboard – a demonstration tape sent to recording studios included melodies performed with the sampled sounds of burping (Beethoven's 'Ode to Joy') and bouncing basketballs (The Beatles' 'When I'm Sixty-Four'). The only customer was Stevie Wonder who used it to reproduce the sounds of birds, bugs, and other non-humans on *Stevie Wonder's Journey Through The Secret Life of Plants* (1979): 'The very first thing Stevie did was take a recording of a single note from a bird he had recorded in Hawaii and use it to play the melody from the second track called 'The First Garden'. It completely blew him away. Me too!' (Mendell 2015). The instrument was, in the words of Coren, 'not a commercial success'. In 1978, the French company Publison launched a Stereo Digital Audio Computer called the DHM 89 B2.[21] It was a digital delay device that was able to repeat memorised sounds indefinitely. By 1980, it could be used in conjunction with a KB 2000 keyboard, which Publison introduced to play back any sound that had been recorded by tape or microphone: 'Put any sort of sound in memory and tame it!' a trade show flier advised. The Fairlight CMI may not have been the first instrument to make 'sampling' available, but it was the most widely used. Fairlight Instruments and their distributors successfully connected the engineering worlds of its designers with the musical worlds of its users.[22]

Analogue/Digital, Accidents, Authenticity

The use of digital technologies since the 1980s has reshaped the processes of musical production. The use of digital technologies has also changed the ways in which music is stored, distributed, and consumed. This book, though, challenges the view that a digital revolution is currently replacing analogue ways of doing things. Instead of accepting arguments found in academic and non-academic writing about a transition from analogue to digital (Warner 2003) or the entering of a 'digital era' (Blake 2007), this book employs empirical evidence to present a more dispassionate approach to the study of digital technologies.[23]

Using the sampler as a case study and focusing on co-existence and continuity as well as change, I show how digital synthesizer/sampling technologies were designed and used in ways that were consistent with older discourses, narratives, and practices. When enough sample time was available in the mid-to-late 1980s, hip-hop producers began sampling and looping sounds from vinyl recordings that had previously been repeated and extended using turntables and magnetic tape. These analogue technologies continued to be used alongside digital samplers. The use of digital sampling technologies that were part of the Fairlight CMI and the Emulator involved learning new technological practices but they were used for doing things that analogue technologies had been used to do in the past. Despite being introduced and marketed as revolutionary instruments that offered users greater creative freedom, the design and use of digital synthesizer/sampling technologies were part of a longer historical process.

In rethinking the history of digital sampling with a focus on its instruments and the technological practices of their users, this book highlights the contingencies in the design and making of musical instruments. Like New England Digital, the histories of companies like Fairlight Instruments and E-mu Systems were shaped by shifting technological developments and commercial priorities. Neither of these companies set out to design a sampler. Michael Kelly, who was joint Managing Director at Syco Systems, the company that distributed the Fairlight CMI in the UK, recalled how Peter Vogel at Fairlight told him: 'God, don't sell it for its sampling. We only put the 'mic in' on the back as a last-minute afterthought!' (Kelly 2015). The contingencies in the process of designing sampling instruments are mirrored in the ways these technologies were adopted by users. While Dave Rossum, Scott Wedge, and Marco Alpert at E-mu encouraged users to re-design the Drumulator drum machine by sampling sounds of their own, Vogel did not expect the Fairlight CMI Series I to be used to sample pre-existing recordings or copyright infringement to become a legal issue around the use of sampling instruments.[24] A compressed version of the first three chapters in the book might read: Fairlight Instruments developed a digital synthesizer to imitate acoustic instruments; users of the Fairlight CMI began to sample 'the sounds of everyday life'. E-mu Systems developed a sampling keyboard and encouraged users to sample 'the sounds of everyday life'; users in hip-hop began to sample drum breaks from pre-existing recordings.[25]

Like the accusation of artifice aimed at new technologies throughout the history of popular music, from the use of microphones and vocoders to more recent software plug-ins like Auto-Tune, digital samplers were associated with fakery and inauthenticity in the 1980s (Frith 1986). However, this book demonstrates that the realism and authenticity of sampled sounds was as important to the designers and users of early sampling technologies as the fidelity of sound was to the designers of early sound reproduction technologies. Emily Thompson and Jonathan Sterne both write about the discourses of fidelity and realism that accompanied the invention of late nineteenth-century recording technologies. In her history of tone tests in the US between 1915 and 1925, Thompson outlines The Edison Phonograph Company's quest for 'phonographic fidelity' (2005, p. 134). Sterne describes how Victor Talking Machine Company used their advertisements to market the way its recording technologies could achieve 'true fidelity' by reproducing sound as a 'vanishing mediator' (2003, p. 283). This concept is also relevant to the use of early digital synthesizer/sampling instruments like the Fairlight CMI to imitate the 'real sounds' of acoustic instruments. Rather than devices associated with artifice, this book outlines how discourses of fidelity and authenticity were central to the design and marketing of sampling technologies like the Fairlight CMI and E-mu Emulator. I also illustrate how an ideology of authenticity is still important for the contemporary users of digital technologies, including virtual instruments like software samplers and synthesizers, and why users in different musical worlds continue to value 'real' sounds, 'real' instruments, and 'real' performances'.

'The Rest is History': Writing a History of Music Technologies and their Users

Instead of exploring the ways samplers have been used as music technologies, the academic literature on digital sampling has tended to focus on issues relating to authorship (Sanjek 1994), copyright (Schumacher 1995, Hesmondhalgh 2006), morality (Porcello 1991), postmodernism (Goodwin 1988), and gender/sexuality (Bradby 1993, Loza 2001). A central argument I make in this book is that it is necessary to study digital sampling within the cultural and social contexts of *music making*: in professional studios, home studios, concert stages, performance spaces, and other sites of musical production. Here, I draw on the work of scholars such as Tara Rodgers who has made a valuable attempt to 'shift the focus from well-worn debates over copyright infringement issues by pointing toward greater understanding of the musical attributes of samplers and other digital instruments' (2003, p. 313). Rodgers, though, chooses to restrict her study of sample-based music to the 'underground', a term that excludes much of popular music and many key users of samplers/sampling instruments. My research for this book began with an attempt to focus on digital sampling within the contexts of popular music making. By the end of my research, I realised that it was not possible to write about the history of sampling, or understand the design and use of sampling instruments, without moving between the worlds of art, folk, *and* popular music.[26]

To provide context for my discussion about the contemporary uses of sampling technologies in the second part of the book, the chapters in the first half contain detailed studies of sampling technologies and their design by Fairlight Instruments (chapter one and two) and E-mu Systems (chapter three). The archival research I carried out about specific instruments by consulting back issues of magazines like *Sound International*, *Studio Sound*, *Sound on Sound*, and *Keyboard* was supplemented by interviews with their engineers and designers. For example, I conducted an email interview with Peter Vogel of Peter Vogel Instruments to establish his 'facts' about the Fairlight CMI – the design objectives, its manufacture and distribution, the sale and marketing of the instrument, the relationship between the designers and users – as it was difficult to rely solely on existing primary and secondary sources containing conflicting information. Longer interviews with the users of early sampling technologies including Richard Burgess, J.J. Jeczalik, and Keith LeBlanc also shape the material in the first three chapters of the book and I conducted these in person or via Skype. These users were asked to recall events that occurred more than thirty years previously. While the answers of some interviewees seemed less reliable because of their need to write themselves into the history of digital sampling and its technologies as innovators/pioneers, others were careful to avoid myth making by admitting how much they were unable to remember.

My aim has been for this book to be an accurate history of digital sampling and its technologies, but it is not a definitive one. As Antti-Ville Kärjä suggests, 'Writing history is always about selecting things to tell – writing total history is impossible' (2006, p. 4). I made choices about which of the early synthesizer/sampling technologies to focus on as the most important in the history of sampling – the Fairlight CMI and E-mu Emulator over the Synclavier, for example. The designers of the Fairlight CMI are quick to claim that they 'changed the way we make music, forever' or refer to 'the classic Fairlight sound that defined the music of the eighties'. Those writing histories about instruments and their designers also tend to draw on a series of myths about the construction of new technologies, including the 'light bulb' moments experienced by 'geniuses' that change history, and musical history, forever.[27] One of these 'eureka' moments can be found in an article celebrating twenty five years of Fairlight Instruments and is discussed in more detail in the first chapter: Peter Vogel accidentally stumbles across the realisation that digitally sampling the sound of a piano playing on the radio sounded better than trying to create a similar sound using digital synthesis. 'The rest is history' writes journalist Rita Street (2000). However, history was still to be written by the users of sampling technologies in ways that were shaped by but completely unforeseen by their designers.

Notes

1 The makers of the Fairlight CMI, Fairlight Instruments, went into receivership in 1988. Peter Vogel, one of the designers of the Fairlight CMI, started Fairlight Instruments Pty Ltd in 2010 before the launch of the CMI-30A. Its name was changed to Peter Vogel Instruments in 2012.

2 Kim Ryrie stated: 'I remember at the AES [Audio Engineering Society] show in 1979, we had people coming up to the instrument, which was being used to demonstrate how you could play natural sounds, and they couldn't understand what it could be used for! It was a case of, 'Yes, that's great – but who can make music with such a thing?'' (quoted in Gilby 1987a, p. 52).

3 Gabriel Rossi Rognoni writes: '[O]rganology is acknowledged as [the] authoritative repository of technical information about musical instruments but regarded as unconcerned with. . .the broader context – music, culture, society' (2017). On the new organology, see Tresch and Dolan 2013.

4 For a useful discussion of technological determinism and theories of technology and society, see Taylor 2001. For an influential introduction to the social study of technology and more on those who began challenging the idea of the heroic inventor, see MacKenzie and Wajcman 1985.

5 The concept of interpretative flexibility is derived from The Empirical Programme of Relativism (EPOR) in the Sociology of Scientific Knowledge (SSK) to refer to the different interpretations of scientific facts before consensus emerges. In applying it to the sociology of technology, Bijker and Pinch use the

safety bicycle as a case study to demonstrate how its development from earlier artifacts like the Boneshaker and the Penny Farthing was multidirectional rather than linear. The stabilisation of the technology as a bicycle with two (or three) wheels of equal size is part of a process that ends with closure or a 'closure mechanism'.

6 See Bernstein and Payne 2008 for an interview with Buchla about his approach to design: 'I was never tempted to build keyboards into synthesizers. To me, that was unnatural' (p. 167). For more on the differences between Buchla's and Moog's synthesizers, see Pinch and Trocco 2002.

7 See articles in newspapers/music blogs with titles such as 'Back to the Future: I'm in the Moog Again' (Gregory 2015). In 'The Synth Revival: Why the Moog is Back in Vogue', Richard Norris of The Grid states: 'I don't want to use a mouse and laptop anymore; writing something in music software has become like drawing on an Etch A Sketch' (quoted in Boxer 2015).

8 In a dictionary of conceptual terms, Roy Shuker defines sampling as 'the practice of using computer technology to take selected extracts from previously recorded works and using them as part of a new work, usually as a background sound to accompany new vocals' (2017, p. 306).

9 For examples of articles that appeared in the UK and US music press in the mid-to-late 1980s about the legal issues around the digital sampling of sounds from pre-existing recordings, see Barry 1987, Gray 1987, Sutcliffe 1987, and Torchia 1987.

10 At Radio Corporation of America (RCA), engineers Harry F. Olson and Herbert Belar began developing electronic music synthesizers in the late 1940s. For more on this, and other early uses of computing technologies to make music, see Manning 2013.

11 For an outline of the principles of digital audio processing and digital synthesis, see Holmes 2016. For more about the work of Nyquist at Bell Laboratories and its relationship with Claude Shannon and Warren Weaver's *Mathematical Theory of Communication*, see Sterne 2012b.

12 Reviews and articles about the Fairlight CMI Series I and II referred to it as a digital synthesizer and music computer, not a sampler. See Crombie 1979, Farber 1980, Levine and Mauchly 1980, Elen 1981, Meredith 1981, Williams, E. 1982, Williams, N. 1982, and Dawson 1983.

13 For more on the history of magnetic tape recording, see Malsky 2003 and Brøvig-Hanssen 2013.

14 The imitation of string, wind, and percussion instruments was a goal of the electronic synthesizer designers at RCA Laboratories in the 1950s: 'The electronic music synthesizer has been used to provide simulations of the voice and existing musical instruments as well as entirely new musical tones which cannot be produced by the voice or existing musical instruments' (Olson and Belar 1955, p. 608). For more on the use of synthesizers to imitate acoustic instruments, see Pressing 1992 and Jenkins 2007.

15 The idea of control appears regularly in the discourse of both designers and users of digital synthesizer/sampling technologies. Tara Rodgers writes about how this relates to gender: '[A]esthetic priorities of rationalistic precision and control epitomize notions of male technical competence and 'hard' mastery in electronic music production' (2010, p. 7).

16 MIDI is an acronym of Musical Instrument Digital Interface. It was developed in the early 1980s to enable instruments using digital technology to be connected together and communicate. Paul Théberge writes that it is 'widely

regarded as one of the most significant innovations in electronic instrument design since the invention of the synthesizer itself' (1997, p. 74).

17 Appleton writes: 'The feature of the instrument that attracted most public attention was its 16-track recorder. The performer could start the recorder, and the computer memorised what was played. Unlike tape, the recording, called a sequence, could be instantly played back at the touch of a button. Tracks could be added on top of each other. . .' (1989, p. 24).

18 Cameron Jones told me how: 'New England Digital introduced the 'Monophonic Sampling' option ('Sample-to-Disk') in 1982. I developed the software during 1981. The Sample-to-Disk option used a 16-bit analogue-to-digital converter that could capture sound at a 50 kHz sampling rate. That was revolutionary since it provided full-fidelity digital sampling' (Jones 2015).

19 The first four compositions produced using the Synclavier by Appleton, Lars-Gunnar Bodin, Russell Pinkston, and William Brunson can be heard on *The Dartmouth Digital Synthesizer* (1976). The sleeve notes ask: 'Why use a digital synthesizer instead of a Moog or other analogue synthesizer? The answer is the ability to create, by digital means alone, time-variant timbres which make all natural sounds interesting to our ears'.

20 Manufactured by Digital Equipment Corporation (DEC), the makers of the PDP-8 minicomputer, it cost $20,000[US]. Coren told me that 'before the product had a name and long before Computer Music Incorporated existed, the product was simply a late 1960s vintage Moog synthesizer attached to an A/D converter and a PDP-8 minicomputer' (Coren 2015).

21 The owner of Publison, Peter Dean, told me: 'This machine was exhibited for the first time at the AES show in Hamburg, Germany (February 28-March 3, 1978). It became famous as it was the only machine to perform high quality pitch-shifting, without audible glitches' (Dean 2015).

22 In his study of the Minimoog and its adoption as an instrument in rock, Trevor Pinch writes: 'It is sellers who tie the world of use to the world of design and manufacture. Sellers are 'boundary shifters'. They are the true 'missing masses' of technology studies' (2003, p. 270).

23 On definitions of analogue, digital, and the relationship between them, see Peters 2016 and Sterne 2016. In an earlier article, Sterne writes: 'Digital technologies are best understood as always bound up with a range of cultural practices and other 'analogue' technologies' (2006, p. 95).

24 Vogel told me: 'We didn't think about copyright at all. We sampled the vast majority of the library sounds ourselves. Some of them were contributed by users. It's a long way short of the sampling that artists do these days, where they take whole phrases. The technology of the day was so limited we were hard pushed to sample more than one note anyway!' (Vogel 2011a).

25 I use the sounds of everyday life as shorthand for a range of music making activities using sound reproduction technologies in ways traditionally associated with film Foleys, *musique concrète*, and field recordings. These include recording the sounds of human/non-human environments and the use of everyday (and non-everyday) objects such as glass and guns to make music. For more on the ambiguity of the everyday in social theory, see Sandywell 2004.

26 Simon Frith draws on Howard Becker's idea of 'art worlds' (1982) and Pierre Bourdieu's concept of 'cultural capital' (1984) to discuss how musical taste and consumption work within these three different musical worlds that should be understood as an interrelated field: 'What is involved here is not the creation and

maintenance of three distinct, autonomous music worlds but, rather, the play of three historically evolving discourses across a single field' (1996, p. 42).

27 Outlining and questioning the 'notion of the heroic inventor', MacKenzie and Wajcman (1985) write that '[g]reat inventions occur when, in a flash of genius, a radically new idea presents itself almost ready-formed in the inventor's mind. This way of thinking is reinforced by popular histories of technology, in which to each device is attached a precise date and a particular man (few indeed are the women in the stereotyped lists) to whom the inspired invention 'belongs'" (pp. 7-8).

PART I

INSTRUMENTS

ONE

Tomorrow's Music Today

The Fairlight CMI Series I and II

My focus in this chapter is the Fairlight Computer Musical Instrument (CMI) and, more specifically, the CMI Series I and II. As outlined in the introduction, its designers at Fairlight Instruments, Peter Vogel and Kim Ryrie, were primarily interested in the use of digital synthesis to replicate the sounds of acoustic instruments. As a result of failure and serendipity in the design process, they ended up using digital samples instead. Unimpressed by the fidelity of its pre-recorded sounds, Richard Burgess used the CMI to record 'the sounds of everyday life', incorporating them into recordings for Kate Bush; composer Peter Howell and other members of the BBC Radiophonic Workshop used its sampling function to combine the sounds of acoustic instruments with random noises, creating new instruments and libraries of sound effects; without an instruction manual, hip-hop producers, Afrika Bambaataa, Arthur Baker, and John Robie, experimented with the pre-set sounds of the sample library, but not to imitate the sounds of acoustic instruments. These are examples of musicians using the instrument in unexpected ways and of users failing to follow, what Madeleine Akrich refers to as, the 'script' inscribed in the technical object. To understand these objects, Akrich writes, '[W]e have to go back and forth continually between the designer and the user, between the designer's projected user and the real user' (1992, pp. 208-209). In this chapter, I draw on archival research and interviews to focus on both the designers and the projected/actual users of the Fairlight CMI. However, I want to start by taking Trevor Pinch and Karen Bijsterveld's advice and begin by 'follow[ing] the instrument[s]' (2004, p. 639).

Following the Instrument: The Fairlight CMI Series I

The Fairlight CMI consisted of a large Central Processing Unit (CPU) with two microprocessors and two 8-inch floppy disk drives, a QWERTY keyboard with a monitor, and two six-octave keyboards (Figure 1.1). There were three ways that users could generate new sounds with the instrument: sampling external sounds, using additive synthesis, or drawing waveforms

Figure 1.1 Fairlight Computer Musical Instrument (CMI) Series I

with a light-pen attached to the monitor. The light-pen was successful with audiences at Audio Engineering Society (AES) conventions where the CMI was demonstrated to potential customers but irked rival companies who rejected it as a gimmick. Cameron Jones of New England Digital (NED) dismissed it by saying that '[u]sing a light-pen to draw a visual representation of a sound wave is kind of like using a pencil to draw a high-resolution JPEG image' (quoted in Milner 2009, p. 317). Roger Linn was blunt:

It was completely useless, a stupid idea, because you're only going to get very odd and bad harmonics, which was emphasised by the fact that [the] Fairlight's sampling rate and bit width was so low. It was a feature they kept talking about, like you could 'make any sound,' but imagine making any sound by drawing a waveform. It's just impossible.

(quoted in Milner pp. 317-318)

While engineers criticised the non-user-friendly interface of a monitor and light-pen, musicians like Peter Gabriel and Herbie Hancock were interested in the opportunity to make music with touch-screen technology. Gabriel described his experiments with it:

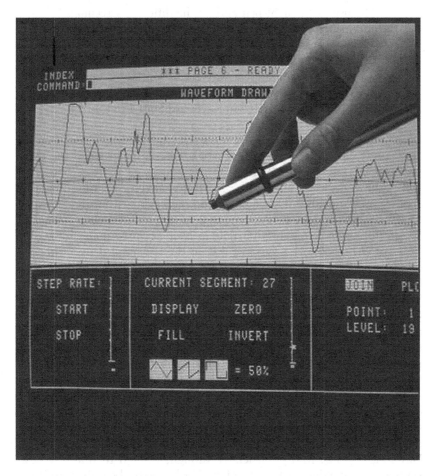

Figure 1.2 Fairlight CMI Control Page 6 (Waveform Drawing) with Light-Pen

> You have a sort of TV screen, which you work with so you're working to some extent with the waveform. You have a light-pen that you can use to programme the waveform on to the screen and you get a considerable amount of control over the sound like that. So, as they develop the visual correspondence I think it will become easier. At the moment it does take a little time to get the hang of it, but I think it's within anyone's grasp if they have the time available.
>
> (quoted in St. Michael 1994, pp. 81–82)

Hancock was effusive with praise about the light-pen and the instrument more generally:

> The use of the light-pen and all the different screenings and menus you have available, and the different ways in which you can manipulate sound, are incredible. The fact that you can draw your waveforms, loop any points you want, and merge different waveforms together is fantastic. There's nothing even close to that as far as I know.
>
> (quoted in Keyboard Staff 1983, p. 53)

Despite this enthusiasm, users did not overcome the difficulties of creating sounds with the light-pen and it was discontinued when the CMI Series III was launched in 1986.

As well as the use of a light-pen to draw waveforms, the sounds of acoustic instruments could be played on the keyboards of the Fairlight CMI using the library of pre-recorded samples stored on one of two 8-inch floppy diskettes. In an article in *New Scientist* magazine, Giles Dawson wrote: 'Insert a systems disc in the left-hand drive, a library disc in the right, and you can explore a world of sound limited only by your imagination' (1983, p. 333). Rather than being used to create 'any sound you can imagine', there were fears about how the Fairlight CMI and other digital synthesizer/sampling instruments would be used to imitate the sounds of acoustic instruments. For this reason, trade unions representing the economic interests of performing musicians did not welcome them.[1] In the UK, the Musicians' Union (MU), which has traditionally campaigned about issues relating to the live performance of music, was concerned with protecting the rights and employment opportunities of its members.[2] Their spokesperson, Maurice Jennings, suggested: 'If you want the sound of violins, book violins; if you want the synthesizer sound, we've no objection to synthesizers' (quoted in Dawson, p. 334). Many musicians, though, wanted to find out how synthesizers like the Fairlight CMI could be used to digitally reproduce the sounds of acoustic instruments like violins and how 'faithfully' they could do so.

Along with the pre-recorded library, external sounds, including the performance of acoustic instruments, could be recorded or 'sampled' using a microphone or line input. Electronic music composer Eberhard Schoener, an early adopter of the Fairlight CMI, was evangelical about the instrument and its use for imitating orchestral sounds:

> The Fairlight is incredible. . .you can make a sound that is just like – snaps fingers – which you can programme and make a whole symphony from. You keep a library of sounds on floppy disc. So, with a Fairlight you can have a Steinway piano sound or whatever you want. You can blend and shape sounds however you wish.
>
> (quoted in Denyer 1980, p. 16)

However, with a sample rate of 24 kHz, it was difficult to reproduce the sounds of acoustic instruments with a level of fidelity that satisfied users who were experts in the field of recorded sound. In a report delivered to the 1980 International Computer Music Conference, audio data consultants, Steve Levine and J. William Mauchly, concluded:

> Steinway needn't worry about competition from this instrument. In general, the Fairlight offers an enormous palette of sounds to the musician, but it can't do everything. Like a camera, the CMI becomes transparent to the viewer, with no characteristic sound of its own.
>
> (1980, p. 566)

Despite the fears of organisations like the MU, the use of the Fairlight CMI to reproduce the sounds of acoustic instruments with a level of fidelity acceptable for some users was not possible. This was because of its technical constraints or, to use James J. Gibson's (1979) term, the 'affordances' of the technology.[3] These restrictions were the result of the high price of microprocessors and the availability of Random Access Memory (RAM). The sample time on the Fairlight CMI Series I also meant the length of sounds that could be digitally recorded and reproduced was limited to one second.

As well as issues over the sound quality of its samples, one of the other reasons users found it difficult to replicate the sounds of acoustic instruments was because the Fairlight CMI was a keyboard-based instrument. It was based on the same way of organising sound as an older technology, the piano.[4] While some users struggled with typing instructions on the QWERTY keyboard, its successful adoption by other users was due to the inclusion of its two piano keyboards.[5] Richard Burgess, who used a Fairlight CMI on the recording sessions for Kate Bush's *Never for Ever* (1980) album, told me about the design interfaces of analogue synthesizers and the problem of using them to generate sounds without a keyboard. His first synthesizer, the EMS Synthi A, was built into a plastic briefcase consisting of aluminum knobs, a matrix, and a joystick-like controller but did not have a touch-plate keyboard like the Synthi Aks.[6] In comparison, the inclusion of piano keyboards with a digital synthesizer like the Fairlight CMI

> . . .made it easy for people to relate to it. I've played keyboards on tons of records but I'm not a great keyboard player by any means. That's one of the reasons why I liked the whole computer thing. I could play the part, programme it, and fix the mistakes. It made it good. For instance, on the Kate Bush sessions I'd programme stuff and I played all the percussive parts but if there was a melodic part someone else would play it. It was obviously a smart move putting a keyboard on the front of it so that it was relatable to anybody.
>
> (Burgess 2011)

Sounds could be played on the Fairlight CMI using the same gestures as other keyboard instruments, but users encountered problems when trying to imitate the sounds of stringed or brass instruments. Stephen Paine, cousin of Peter Gabriel and co-owner of Syco Systems, recalled: 'It became clear after a while that it was impossible to achieve the expressiveness with a keyboard that players of acoustic instruments have with finger and/or mouth control' (quoted in Tingen 1996a, p. 50). Using the Fairlight CMI to replicate the sounds of orchestral instruments involved practical difficulties that resulted in users experimenting with it in ways that had not been envisaged by its designers.

Following the Designers (and Distributors): Peter Vogel, Kim Ryrie, and Bruce Jackson

Few could have predicted that the first commercially available digital synthesizer/sampling instrument would be designed in Australia. Burgess was involved in the development of drum synthesizers and electronic drum kits like the Simmons SDS-5 in the late 1970s, as well as being an early user of the Fairlight CMI:

> You could not have expected that anyone would come up with a device like that at that time, I don't think, and certainly not from Australia. You would have thought

it would have come out of one of the major universities like MIT or Stanford or IRCAM or somewhere like that.

(Burgess 2011)

The development of digital synthesizer/sampling technologies might also have been expected to occur first in the Far East.[7] However, as Ralph Denyer described in an article in *Sound International* in May 1980, Fairlight Instruments had an advantage:

> Although several manufacturing companies have built prototype digital synthesizers and circulated technical data at exhibitions, the Australian Fairlight company have pipped the Japanese, Europeans, *and* [original emphasis] the Americans to the post by getting their instrument into production first.
>
> (p. 16)

While the designers of the Synclavier and the Computer Music Melodian worked in the engineering department at Dartmouth College and the electronic music studio at University of Pennsylvania respectively, Fairlight Instruments grew from a looser arrangement of social and family networks. The initial design of the Fairlight CMI was the result of engineering experiments in makeshift spaces in the suburbs rather than university departments or the laboratories of research institutes. Ryrie and Vogel were not university graduates but high school friends who bonded over building electronics. As they developed their synthesizer, an engineer who *had* received the support of both a higher education institution and a government-funded scheme joined the company.[8]

Ryrie and Vogel started Fairlight Instruments in 1975. The company was named after a hydrofoil that passed Point Piper, a suburb of Sydney, where they had set up a technology laboratory in the basement of Ryrie's grandmother's house. For Vogel, designing a digital synthesizer

> . . .was inspired by the modular Moog, which could make amazing music but was very finicky to set up and only monophonic. Microprocessors were just becoming available in the 1970s and I thought there must be a way to use them to eliminate all the knobs and patch leads of the usual synth.
>
> (Vogel 2011a)

Moving to new premises in 1976, Ryrie and Vogel began working with Tony Furse, a microprocessor technology expert, on prototypes of the Fairlight CMI.[9] They saw the electronic emulation of acoustic instruments using digital synthesis as 'the holy grail' (*ibid.*). However, the difficulties involved in doing so led them to digitally record (or sample) sounds to analyse their waveforms. Vogel explained what happened next:

> On a whim I decided to see what would happen if I changed the software to allow the sampled sound to be replayed at a pitch determined by the keyboard. The sampling time was less than a second and the first source connected to the ADC [analogue to digital converter] was a radio I had going with some music playing. I captured a fragment of a piano note and when I played it back on the keyboard I was surprised how good it sounded, especially polyphonically.
>
> (quoted in Street 2000)

Despite this breakthrough in their attempts to imitate the sounds of acoustic instruments, Ryrie was unhappy at only being able to do so using sampling rather than synthesis:

> We wanted to digitally create sounds that were very similar to acoustic musical instruments, and that had the same amount of control as a player of an acoustic instrument has over his or her instrument. Sampling gave us the complexity of sound that we had failed to create digitally, but not the control we were looking for. We could only control things like the attack, sustain, vibrato, and decay of a sample, and this was a very, very severe limitation of the original goal that we had set ourselves. We regarded using recorded real-life sounds as a compromise – as cheating – and we didn't feel particularly proud of it.
>
> (quoted in Tingen 1996a, pp. 48–49)

The importance of control, fidelity, and other claims in advertising materials aimed at its projected users, shows how the aims of the Fairlight CMI designers were based around the discourse of technological progress and the extension of creative freedom. As Paul Théberge (1997) points out, these have been recurring themes in the marketing of musical instruments. From pianos in the nineteenth century to electric instruments in the twentieth, there has been an emphasis on having more control over a greater range of sounds.

There was a tension in the marketing of the Fairlight CMI between the original goal of using a digital synthesizer to imitate the sounds of acoustic instruments and the presentation of a new, revolutionary instrument. In September 1980, an advertisement in *Sound International* described the Fairlight CMI as 'an entirely new concept in electronic musical instruments' (Fairlight 1980) with the ability to 'sample 'natural' sounds from, say, a microphone, which can then be played on the keyboard and manipulated in various ways' (*ibid.*).[10] A later advertisement for the Fairlight CMI in the February 1982 issue of *Keyboard* magazine asked the question 'Orchestra for sale?' (Figure 1.3), as if the fidelity of sampled sounds was making orchestral musicians redundant. By 1983, Fairlight Instruments had launched the CMI Series IIx and were promoting how users could overcome the limitations of the imagination to produce sounds that had not yet been conceived, rather than the realism and authenticity of the instrument's sounds:

Figure 1.3 'Orchestra for sale?' advertisement (*Keyboard*, February 1982)

> This is the story about a new concept in music production. It goes well beyond the ideas of musical instruments as we know them. It is a concept inspired by the wish to create literally ANY type of music, no matter how complex or difficult to express. To incorporate literally ANY type of sound – not only classical and modern instruments but sounds of the world.
>
> (Fairlight 1983a, emphasis in original)

The removal of claims about the realism of sounds may have been a result of the technology being used by *actual users* in ways that did not correlate with the projected uses imagined by the designers. On the concept of sound fidelity that developed at the start of the twentieth century, Sterne writes: 'Sounds could neither hold faith nor be faithful – that task was left to listeners and performers' (2003, p. 282). When instruments like the Fairlight CMI were being used to digitally reproduce sounds at the end of the twentieth century, the test of fidelity was left to the users.

To connect the worlds of design with the musical worlds of users, individuals and companies in different territories undertook the role of marketing and selling the Fairlight CMI. Syco Systems were the sole agents in the UK. In the US, one person sold the instrument initially. Bruce Jackson had lived next door to Ryrie's grandmother in Point Piper and was an old friend of Ryrie and Vogel. While the latter did not have any contacts in the music industries, Jackson had been working in the US as a live sound engineer:

> I helped them by setting up distribution in the US, where I was friends with a whole bunch of musicians like Rick Wakeman and Tony Bongiovi. . .Tony was kind enough to let me set it up in an unfinished studio he had. I flew all over the US promoting Fairlight in my private plane for almost a year before anyone bought one. Remember, this was the first music sampler ever made, so it wasn't as if people were automatically into it – the concept was entirely foreign at this stage. I remember taking it into Power Station with Bruce Springsteen and he said: 'Ah yeah, BJ that's great, but what am I gonna do with it?'
>
> (quoted in Stewart 2005, pp. 67–68)

In the early 1970s, David Van Koevering travelled around the US to develop a dealer network and a market for the Minimoog because music instrument retailers had never sold synthesizers (Pinch 2003). At the end of the 1970s, Jackson flew across the country to demonstrate the Fairlight CMI at the homes of individuals like Herbie Hancock and Stevie Wonder.[11]

Van Koevering was trying to persuade amateur and semi-professional rock musicians to buy a smaller, portable, keyboard version of Moog's modular synthesizers for $1,500 [US]; Jackson was trying to sell an expensive digital synthesizer and computer musical instrument costing over $25,000 [US] to an elite group of musicians and recording studio owners. Unlike Van Koevering, he was dealing only with the end users.

The Users

One of the earliest advertising slogans that accompanied the Fairlight CMI in the 1980s was 'Tomorrow's Music Today'. A full-page advert in the September 1980 issue of *Sound International* warned readers: 'Turn this page and the future of music is passed'. The opportunity to imagine what the future might sound like if programmed and played on a Fairlight CMI was restricted by its expense. Figures ranging from £12,000 to £27,500 for the Series I can be found online and have become part of the mythology of the machine – the more accurate figure is £13,000 (Crombie 1979). The price of the Fairlight CMI may have limited the number of owners who could afford one, but publicly funded institutions also bought them, which meant that its users were more diverse. The technology was not restricted to the world of music and began to be used in education, health, and public service broadcasting. Simon Emmerson, who lectured in the Applied Arts Department of the City University, London, described how 'our postgraduate students use the hardware in ways no one else has before' (quoted in Dawson 1983, p. 335). Geoff Twigg, who taught composition at Goldsmith's College in London, was employed by local education authorities to help children with learning difficulties use the Fairlight CMI and another digital synthesizer, the Syntauri alphaSyntauri.[12] Giles Dawson's article in *New Scientist* also documented how the Fairlight CMI was being used to improve computer literacies. Users in a variety of social worlds, including those in the public sector and private companies, were making music with the instrument and learning how it could be used.

I want to present three short case studies that illustrate how the Fairlight CMI Series I and II was being used in the worlds of electronic and popular music in the late 1970s and early 1980s. Firstly, I draw on primary sources including interview material to examine its use as part of the practices of progressive rock and pop musicians: Peter Gabriel, who was assisted by Peter Vogel, and Kate Bush, who was assisted by Richard Burgess and John Walters in EMI's Abbey Road Studios. I then use secondary sources including histories of the BBC Radiophonic Workshop to examine its use by employees/composers like Peter Howell and Roger Limb. Lastly, I draw on secondary sources about the production of electro and hip-hop in a Manhattan recording studio to investigate its use by Afrika Bambaataa, Arthur Baker, and John Robie in the making of 'Planet Rock'. The uses of the Fairlight CMI I want to explore can be categorised under three headings: (i) replicating 'real' or 'natural' sounds, or sampling 'the sounds of everyday life'; (ii) creating 'new' sounds; and (iii) experimenting with sounds from the Fairlight CMI's sample library. In the initial stages of its adoption by musicians, artists such as Herbie Hancock and Stevie Wonder used the Fairlight CMI as part of their live performances on stage. I will, however, mainly be examining its use in a series of professional recording studio spaces.

(i) A Social Network: Richard Burgess, Peter Gabriel, and Kate Bush

Outwith the context of large institutions like the BBC and university departments, the social networks of freelance/professional musicians are important for understanding who was using the Fairlight CMI in the early 1980s. Richard Burgess is one of the key users and intermediaries in this story, describing himself as a 'studio musician playing sessions' (Burgess 2011). Inspired by forecasts about the future in Christopher Evans' *The Mighty Micro* (1979), he developed an interest in using microprocessor-based technologies. These included the Roland MC-8 MicroComposer (1977), a digital sequencer Burgess used to programme individual parts on the Landscape album *From the Tea-rooms of Mars . . .* (1981) and its single, 'Einstein A Go-Go'.[13] His work on the development of the Simmons SDS-5 electronic drum kit had shown him the limitations of using analogue synthesis to re-create the sounds of acoustic drums:

> You realise that no matter how many oscillators you have and how many times you can add on a harmonic, somehow you never quite get to the complexity of a natural sound. So, when you see the possibility of starting with a natural sound, that's very attractive. You start with complexity rather than starting with simplicity and trying to build complexity.
>
> (Burgess 2011)

Using a Fairlight CMI to digitally sample the sounds of acoustic instruments offered him the kind of realism that was the initial aim of Ryrie and Vogel: 'With sampling you could sample a timpani and it really was a timpani. It really sounded like a timpani' (*ibid.*). Burgess is a little unclear about exactly how or where he first became aware of the Fairlight CMI. It may have been a demonstration at Morgan Studios in Willesden Green, a visit to the Fairlight Instruments office in Sydney, or a trip to the village of Box in Wiltshire: 'I think I heard about it first, then went out to Peter

Gabriel's place out in Box. This was before he had Real World Studios and he was recording the third solo album. I remember going out there and I think that was the first time I ever saw one' (*ibid.*).

One of the first commercially available recordings to feature the sampled sounds of the Fairlight CMI was Peter Gabriel's third untitled solo album – more commonly known as *Melt* – released in May 1980. The recording sessions took place the previous year and Peter Vogel is credited with duties relating to Computer Musical Instrument. When asked about his role on the album, Vogel told me: 'I was staying with Peter Gabriel while he was recording *Melt* and gave him some tuition on use of the CMI, which he had bought. We recorded some tracks that ended up on that album' (2011a). Gabriel began to use it by recording and experimenting with the sounds of milk bottles being smashed and bricks being banged together. His biographer, Spencer Bright (2000), suggests that these can be heard as the track 'I Don't Remember' fades out, though any samples of these everyday sounds do not appear to be audible. Franco Fabbri speculates that the Fairlight CMI may have been used for the string sounds on 'Start' or the bagpipes in 'Biko' but concludes: '[A]ll of these (except perhaps for the bagpipes) could also be generated at that time by analogue polyphonic synths' (2010, p. 179). According to the sleeve notes, the bagpipes are synthesized. In this case, Gabriel and Vogel were the using the Fairlight CMI as a digital synthesizer to imitate the sounds of acoustic instruments and is an example of the technology being used in accordance with the original design objectives. This is not surprising as Peter Vogel of Fairlight was the one programming the CMI: the user of the musical instrument was also its designer.

The results of using the Fairlight CMI to sample 'the sounds of everyday life' are more obvious on Kate Bush's album *Never for Ever*. Released in September 1980, she was assisted by Richard Burgess who had spent time learning to programme the instrument and demonstrate it to prospective users. He explained how his relationship with Syco Systems led him to use the Fairlight CMI on recording sessions with Bush:

> I struck a deal with Syco Systems and demonstrated it to a lot of people. That's how the Kate Bush sessions came about because Kate was friends with Peter [Gabriel]. She called Peter to see if Peter could do the session and I guess he couldn't so they called me and asked if I could do the session with them. So, John Walters and myself threw the Fairlight into the back of my BMW and drove it up to Abbey Road Studios where we did the sessions.
>
> (Burgess 2011)

The sound of breaking glass that Gabriel had been playing with punctuates the first song, 'Babooshka'. Burgess explains the process of recording and playing back the sounds:

> We took glasses, I guess, from the kitchen. We had, I seem to remember, a concrete block or something in the studio and we just threw them down on the concrete block and recorded it. We had several samples and we stacked them up and then just found a combination of keys that made the best sound. The pitch changing is all from the keyboard on the Fairlight and mostly they were clusters, semi-tone clusters, on the keyboard.
>
> (*ibid.*)

The collaborative use of the Fairlight CMI by Richard Burgess and Kate Bush on the *Never for Ever* sessions continued a tradition of introducing everyday sounds into popular music. In the 1960s and 1970s, these were often used to emphasise a particular lyrical theme, or the use of unusual/unexpected sounds created an experience of incongruous juxtaposition.[14] Echoing the founders of *musique concrète*, Bush stated: 'What really gets me about the Fairlight is that any sound becomes music. You can actually control any sound that you want by sampling it and then playing it. Obviously, it doesn't always sound great, but the amount of potential exploration that you have there with sounds is never-ending' (quoted in Diliberto 1985, p. 60). Rather than using the Fairlight CMI to imitate the sounds of acoustic instruments, Bush celebrated the limitless possibilities of the instrument with words that might have been taken from one of its advertisements.

In an approach that would have pleased representatives of the Musicians' Union, Bush did not want the Fairlight CMI to replace the role of performer. She was keen to ensure that performing musicians remained an integral part of the production process:

> I don't feel that I want to create the world's greatest cellist on the Fairlight. I'd rather get a really good cello player in and record him with a good engineer and then use the Fairlight to do something that complemented that. The most exciting thing for me is the combination of real and natural sounds and extremely electronic synthesized ones.
>
> (quoted in Diliberto, p. 72)

Musicians used to working in more traditional ways were puzzled by forms of technological experimentation that seemed illogical: 'She [Bush] had recorded this penny whistle which Paddy [Bush] could play and then played it on the keyboard, and I thought it was a bit of a strange circle. 'Why not just play the penny whistle?!" (quoted in Thomson 2010, p. 166). The answer to keyboardist Max Middleton's question is that the sampling of these acoustic instruments enabled a layering of sounds. Burgess recalls that it added additional textures to songs. Bush's brother, Paddy, provided the raw materials:

> He plays a lot of instruments and he had all these traditional Irish instruments or maybe it was a mandolin and violin and things like that. I can't exactly remember but we sampled a bunch of those and some of those sounds are on 'Babooshka' and 'Army Dreamers' as well. They sound almost Mellotron-ish. The Fairlight [CMI Series I] was 8-bit so it wasn't really high-quality sound so it had a Mellotron-ish quality if you were doing strings.
> (Burgess 2011)

Despite the claims in advertisements for the Fairlight, Burgess refers to a lack of realism in reproducing the sounds of acoustic instruments. In this case, they sound as if they were produced by an older tape-based instrument, a Mellotron. Digital synthesizer/sampling technologies like the Fairlight CMI, which were described as the sound of the future, now sound to this user like an analogue instrument first manufactured in the 1960s.[15]

With a Mellotron, the pre-recorded sounds of acoustic instruments were reproduced using magnetic tape. On the Fairlight CMI, any sound could be recorded and stored digitally. Users then organised these sounds melodically or rhythmically using the six-octave keyboards. The song 'Army Dreamers', from *Never for Ever*, incorporates the sounds of cocking rifles to emphasise the anti-military theme of wasted lives. Burgess explained where the guns came from and expanded on how users of the Fairlight CMI could organise and perform these sounds using a piano keyboard device in ways that were not possible using the older technologies of magnetic tape, Sellotape, and scissors:

> I think the older brother had an arsenal of guns. He brought in a bunch of guns and we tried them all, cocking them, and recording them. . .So we sampled, I don't remember how many guns but quite a lot. . .It was amazing and actually it wound up being multiple weapons on top of each other so it gave it a much more substantial sound because. . .the real thing often doesn't sound like the real thing. That happens a lot in movies. If you want running water, it doesn't necessarily sound like running water if you just record a stream. Sometimes you have to fake it up in order for it to sound correct.
> (Burgess 2011)

Burgess and Bush layered sounds so that the gun noises stored in the Fairlight CMI sounded 'real'; they achieved 'authentic' sounds by using processes that might be described as 'fake' or 'artificial'. As Bush assumed more control of the production process on this album and its follow-up, *The Dreaming* (1982), the perception of her as a pioneering user of the Fairlight CMI contributed to her move away from the stereotype of the teenage pop prodigy that accompanied the reception of her first two albums. She was hailed in one music technology magazine as 'a vital and innovative composer, singer, keyboardist, and producer who has shaped a uniquely personal and organic sound' (Diliberto 1985, p. 57). However, the 'organic sound' and sounds in Bush's recordings were the result of using highly sophisticated, computer-based instruments like the Fairlight CMI with its 8-bit microprocessors, digital synthesis, and sampling technologies.

Along with the sounds of guns and glass, other songs from *Never For Ever* contain the noises of the non-human world. Buzzing insects can be heard on 'Delius (Song of Summer)'. On 'All We Ever Look For', there is a short interlude towards the end of the song. It contains the sounds of footsteps on stairs before the opening of doors that expose bird sounds as well as Hare Krishna chants and the muted clapping of an audience. As with Gabriel's recordings, it is difficult to be certain if these sounds were digitally recorded. Due to problems operating the Fairlight CMI during the sessions, engineer Jon Kelly used the tape-based technique of 'flying in' to insert some sounds into recordings.[16] A biographer of Bush, Graeme Thomson, states:

> Because of the technical limitations of this new machine, several of the sounds that might at first appear to be samples – Hare Krishna chants, countryside noises, random spoken voices – were actually flown in by Kelly using a tape recorder, which at the time gave a much better sound quality.
>
> (2010, p. 165)

As analogue recording technologies continued to be used alongside new sampling technologies, the important thing for Bush was that the Fairlight CMI enabled her to reproduce 'real sounds': 'What attracts me to the Fairlight is its ability to create very human, animal, emotional sounds that don't actually sound like a machine' (quoted in Anon 1982, p. 46). For this user, and at this time, the Fairlight CMI could be used to digitally reproduce the sounds of humans and other animals as a 'vanishing mediator' (Sterne 2003, p. 283), without reproducing the sounds of its own digital production.

(ii) An Institution: The BBC Radiophonic Workshop

From 1958–1998, the BBC Radiophonic Workshop operated as a tape-based and electronic music studio to provide sound effects and soundtracks for television and radio programmes produced by the British Broadcasting Corporation (BBC).[17] Most famous for the signature theme to *Doctor Who* (1963) and the work of female composers like Daphne Oram and Delia Derbyshire, there was, in the 1960s, what Louis Niebur calls, 'a gradual shift away from exclusively tape-manipulated techniques towards the use of sounds produced electronically, first by simple oscillators and then, at the end of the decade, by voltage-controlled synthesizers' (2010, p. 121). The BBC purchased analogue synthesizers in the 1970s including two EMS VCS3s, a Synthi 100 (referred to as Delaware), and an ARP Odyssey. These were welcomed as time-saving devices for the making of sound effects.[18] In 1981, they bought a Fairlight CMI, which was praised two years later in a book that provides an insight into the musical practices of its composers:

> Behind many of the doors, late in the evening, the sounds still continue. Some of them being made, perhaps, on a machine called the Fairlight Computer Musical Instrument, one of the Workshop's most powerful allies to date. Long past are the 'Glowpot Days' of do-it-yourself equipment. Synthesizers are standard aids and have done away with much of the drudgery of realisation. The Fairlight offers an almost alchemical combination of concrete music and electronic music.
>
> (Briscoe and Curtis-Bramwell 1983, p. 56)

Its users at the BBC became less reliant on the mixing desk known as the 'Glowpot' and avoided the difficulties of cutting and splicing tape to create sound effects. These included some of the stranger sounds of everyday life such as germs eating plaque on teeth. One of the Workshop's composers, Roger Limb, explained how it was made:

> I scrunched an apple and put it on tape, then fed it into the Fairlight and started playing it on the keyboard. It worked very well as an effect. There was a sound there that *could* [original emphasis] be arranged in a musical fashion.
>
> (p. 59)

Instead of using the Fairlight CMI to record the sound of eating an apple, Limb was using the older technology of magnetic tape as well. Rather than reproducing the sounds of acoustic instruments, as Vogel and Ryrie had in mind when designing the instrument, these were the types of sounds – 'ANY type of sound' – that were generated by users.

As well as assisting with the construction of unusual sounds, the Fairlight CMI was used in the Workshop to create new sounds and textures by mixing the sounds of acoustic instruments with other noises. Composer Peter Howell is described as having built

> . . .a battery of composite sounds and named them according to their components. Clarjang is made from a clarinet sound combined with a metallic jangle. Pluckvox combines the plucking of a mandolin note with the second half, his own voice.
>
> (pp. 98–99)

Howell praises the Fairlight CMI as another step in the onward march of technological progress. In a statement that could have been lifted from the manufacturer's literature, he describes its creative possibilities as endless: 'With just the Fairlight there are apparently no limits. The road goes on forever' (p. 99). However, adopting the Fairlight CMI was not the next stage in a linear path. The use of sampling technologies to record, store, and playback 'real sounds' provided a solution to problems that had been experienced after the introduction of analogue synthesizers. Co-founder Desmond Briscoe stated:

> In the past, electronic sound has tended to be dehumanised, and boring, because it was created from very basic waveforms. Natural sounds have much more information in them. They are warmer and more interesting than synthesized sounds; using the Fairlight's ability to provide the composer with a means of *playing* [original emphasis] real sounds is a return to the early days without all the disadvantages of tape manipulation.
>
> (p. 57)

Rather than a revolutionary instrument that could be used to create the sounds of the future, users of the Fairlight CMI at the Radiophonic Workshop were recovering the texture of sounds that had been lost as a result of technological changes in the past.

(iii) A Recording: 'Planet Rock' and the Story of ORCH2

'Planet Rock' (1982) by Afrika Bambaataa and The Soul Sonic Force is a recording based on the appropriation of sounds from other recordings: 'Trans-Europe Express' and 'Numbers' by Kraftwerk are its basic building blocks. While Bambaataa, Arthur Baker, and John Robie used the Fairlight CMI on 'Planet Rock', they did not use it to sample sounds from pre-existing recordings. They reproduced the elements of the Kraftwerk recordings using other instruments. The melody from 'Trans-Europe Express' was copied using Robie's keyboards – a Micromoog and Prophet 5 – and the rhythmic pattern of 'Numbers' was reconstructed using a Roland TR-808 drum machine.[19] While Burgess and Bush used the Fairlight CMI to incorporate found sounds into recordings and composers in the BBC Radiophonic Workshop used it to create new sounds from unusual juxtapositions, its use on 'Planet Rock' was different. Despite not having access to an instruction manual or technical support, its producers used the technology in a way more closely aligned with the design objectives of Vogel and Ryrie: they used the sounds of orchestral instruments in the pre-recorded sample library.

Without the luxury of a manual, Robert Fink writes that, "Planet Rock' was pure serendipity. Bambaataa and Baker had no idea how to use the machine, no one to show them, nor any time to learn' (2005, p. 344). They did not know that external sounds could be sampled using the Fairlight CMI and instead sought out sounds from the library of pre-recorded samples. Continuing the theme of contingency, Baker describes how:

> There were a lot of happy accidents when we were making these kinds of records. Like the orchestra hit. We were going through the sounds on the Fairlight, which, although it was worth over $100,000 back then, probably only had what a $1,000 computer can do these days. You couldn't sample on the Fairlight, it was all pre-sampled sounds, so we used an explosion, the handclaps, and the orchestra.
> (quoted in Buskin 2008, p. 82)

Without the availability of advice from the instrument's designers in Australia or those with close connections to the designers and distributors who had taken the time to learn how to use it, no one knew how to use it to sample external sounds.[20] The sounds of the Fairlight CMI on 'Planet Rock' were, however, samples from a pre-existing recording: a recording that had been sampled by the designers at Fairlight Instruments.

The orchestral sound that Baker describes using on 'Planet Rock', the ORCH2, is from a recording of *The Firebird* ballet by Igor Stravinsky. The orchestral sample was digitised by a computer programmer/musician and added to the sample library on the floppy diskettes of the Fairlight CMI. When Peter Vogel Instruments developed a Fairlight CMI mobile application (or app) for iPhone and iPad, which was released in March 2011, it contained information about the sample that aimed to correct a misconception:

> Quite possibly the most ubiquitous and instantly familiar Fairlight sample was the ORCH2 orchestra stab that appeared on more pieces of music than one would care to remember. From this moment on, no self-respecting synth or sampler would be without an orchestra stab patch that was some variation of the ORCH2 sample. This is often misidentified as ORCH5.
>
> (Fairlight 2011b)

The ORCH2 sound was popular in hip-hop and one of the reasons why is because it resembled the sound of scratching with turntable styluses by DJs like Grand Wizzard Theodore and Grandmaster Flash.[21] David Toop describes how:

> . . .the resulting noise, a tearing jolt of electricity, rocketed hip-hop into a new dimension. The effect combined the qualities of a Grandmaster Flash scratch, amplified to monstrous bandwidth, with the science-fiction suggestion of ten orchestras, all playing a single chord in perfect synchronisation.
>
> (2000, p. 99)

Bambaataa, Baker, and Robie used the orchestral sounds in the pre-recorded sample library but were not trying to replicate the sound of acoustic instruments. Listeners like Toop heard the ORCH2 sample as a more powerful digital version of sounds that were part of the existing aesthetic and practices of hip-hop.

The ORCH2 sound also became common in the production of pop music in the early–mid 1980s. In 1985, Kate Bush commented on its overuse in an interview with *Keyboard* magazine: 'Some of the pre-sets that they supply are actually quite good. But there's one favourite that everyone is using, called ORCH5 or something. Every time anyone who has a Fairlight hears it they go, 'Oh no! Not again!" (quoted in Diliberto 1985, p. 64). On an online forum for Fairlight CMI users, Peter Vogel explained its origins:

> Here I am with the very record from which possibly the most famous orchestra stab of all time came off. It was sampled when I was demonstrating the CMI to English musician David Vorhaus at the Fairlight factory, 15 Boundary Street, Rushcutters Bay, around 1978. David wanted to try out the sampling and I grabbed a random record from a nearby box. The rest is history.
>
> (Vogel 2011b)

The history of the Fairlight CMI, then, is one of contingency. Vogel and Vorhaus did not choose the recording of *The Firebird* because they wanted a particular orchestral sound. The sample library could easily have contained the sounds of another recording. The orchestral hit and keyboard stab that became known as ORCH2, and has been mistakenly referred to as ORCH5, might not have become an important part of musical history had Bambaataa, Baker, and Robie not found it as they searched through the library of pre-sampled sounds on diskette. The arbitrary decisions made by the designers of the Fairlight CMI were followed by the serendipitous discoveries of its users.

Notes

1 For a discussion about the uses of electronic music technologies for emulating acoustic instruments from the 1930s onwards and the impact on performing musicians, see Doerschuk 1983. For more about restrictions placed on the use of Moog synthesizers by the American Federation of Musicians (AFM), see Pinch and Trocco 2002, pp. 148–149.

2 For more on the Musicians' Union and, what its General Secretary John Smith has described as, 'the *continual* battle [original emphasis] with technology', see Williamson and Cloonan 2016.

3 Ian Hutchby writes: 'The affordances of an artefact are not things which impose themselves upon humans' actions with, around, or via that artefact. But they do set limits on what it is *possible* [original emphasis] to do with, around, or via the artefact' (2001, p. 453).

4 Paul Théberge's description of the piano as 'one of the dominant musical and cultural forces in the West – theoretically, practically, and symbolically – during the past two centuries' (1997, p. 19) helps explain why keyboards were an integral feature in the design of the Fairlight CMI.

5 Tara Brabazon argues that the Fairlight CMI was popular with users because it combined *both* interfaces: 'One of the reasons for its success was that it applied the design lesson of the Hammond organ and Moog by developing a conventional piano keyboard and a computer keyboard. This dual interface predicted the future of both computing and music, combining their two histories into a package and a meshed future' (2012, p. 101).

6 The Synthi A was a portable version of the EMS VCS3, which as Mark Vail explains, contains 'a tiny patchboard matrix into which you insert pins to route audio and control signals through the device' (2014, p. 88). For more on the Synthi A and Aks, see Jenkins 2007, Pinch and Trocco 2002, and Vail 2000a.

7 As Gene Gregory writes: 'By 1970, East Asia had become the epicentre of the world consumer electronics industry, with Japan in undisputed leadership; virtually all the revolutionary innovations in consumer electronic products since the first transistor radio have come from Japanese industry (1985, p. 7).

8 I am keen to avoid what Tara Rodgers has identified as a tendency in the historiography of synthesizers to venerate white, middle-class, male inventors as archetypal heroes, 'the humble hobbyist and tinkerer[s]' (2015, p. 15) who work in basements to overcome the constraints of the bureaucratic workplace or the family home and 'revolutionise' the processes of music making.

9 Furse was an electronics engineer who set up a company called Creative Strategies Pty Ltd in 1972 and developed hybrid analogue/digital synthesizers called Qasar I and Qasar II. The composer Don Banks, who had set up an electronic music studio at the Canberra School of Music, supported his work. With the help of a grant secured by Banks from the Australian Council for the Arts, Furse began developing a digital synthesizer called the Qasar M8 (Multimode 8), which incorporated microprocessors from Motorola and a monitor with light-pen. A version of the Qasar M8 called the M8 CMI was redesigned at Fairlight Instruments and became a prototype of the Fairlight CMI. For more on Furse's archive, see Chapman 2012.

10 When New England Digital launched the Synclavier II in the same year, they claimed superiority in the imitation of acoustic or 'real' instruments: 'Synclavier II is a revolutionary advancement in synthesizer technology. Its patented digital method transcends 'realism'. Many of its sounds *are real* [original emphasis], virtually undetectable from real instruments. The violins and cellos are so true, you can hear the rosin on their bowstrings' (Synclavier 1980).

11 Hancock used the Fairlight CMI on the hit single 'Rockit' (1983) and demonstrated it on the children's TV programme, *Sesame Street*. Using his thumbprint, Wonder signed a purchase agreement with Fairlight Instruments on 20 November 1979. As well as its use for live performance, the first recording on which he used the CMI was 'Happy Birthday' from the album *Hotter Than July*, released in September 1980. The cost of the Fairlight CMI was $25,220 [US] plus $1,700 for computer crating, airfreight, and customs charges.

12 The alphaSyntauri (1980) was described as an 'affordable' digital synthesizer. For more, see Kellner, Lapham, and Spiegel 1980, Levine and Mauchly 1981, Moog 1981, Acerra 1983, Lehrman 1983, and Greenwald and Burger 2000.

13 Burgess writes: 'We could cut, paste, and copy. We programmed parts that were impossible to play, and changing keys, tempi, sounds, notes, or timings, after having recorded all the parts into the MC-8, was no problem' (2014, p. 138). For histories of Roland, see Kakehashi 2002 and Reid 2004a, 2004b, 2005a, 2005b. On the MC-8, see Hammond 1983, Vail 1990, and Carter 1997.

14 See Pouncey 2002 for more on the relationship between *musique concrète* and rock/pop. He refers to the 'sampling' of traffic noises on 'Summer in the City' by The Lovin Spoonful, jet engines on 'The Letter' by The Box Tops, barking dogs on 'Caroline, No' by The Beach Boys, and bomb blasts on Love's 'Seven and Seven Is'.

15 Burgess's memories of using the Fairlight CMI in the 1980s are mediated by more recent listening experiences: 'I heard a bit of 'Army Dreamers' the other day because it was on YouTube. There's a little flutish sound, somewhere between a flute and a string sound, kinda like the way Mellotrons are. You can't quite tell [but] I'm pretty sure that was the Fairlight' (2011). For more on the Mellotron, see Vail 2000d, Samagaio 2002, Reid 2002, and Awde 2008.

16 The process of 'flying in' was a common way of overdubbing sounds onto multitrack recordings prior to the introduction of Digital Audio Workstations (DAWs). Burgess explains: '[I]t might entail bouncing a segment such as, say, chorus background vocals or handclaps onto a second tape machine and then back onto the master in the desired location' (2014, p. 138).

17 In May 2012, the BBC Radiophonic Workshop was re-established as The New Radiophonic Workshop (NRW) by the BBC in partnership with The Arts Council of England.

18 An internal BBC memo stated: 'One music cue for *Doctor Who* would have taken at least a day to realise by conventional methods [magnetic tape]. With a

mini synthesizer the fastest time was one hour 15 minutes. With the Delaware, after only a few weeks, realisations, more ambitious than ever before were being completed in 34 minutes' (quoted in Niebur 2010, p. 137).

19 The Roland TR-808 was sourced through an advertisement in the *Village Voice* and its owner paid $20 (Brewster and Broughton 1999), $25 (Barr 1998), or $30 (Buskin 2008) for its use on the session. For more on the Roland TR-808, see Vail 1994 and Mansfield 2013. For more about the development of drum machines and their use of microprocessors, see Hammond 1983.

20 The relationship Burgess had developed with the designers and distributors of the Fairlight CMI meant he was able to make a telephone call to Fairlight Instruments in Australia after encountering basic problems such as turning the instrument on.

21 For more on the history of scratching and the role of the DJ in hip-hop, see Poschardt 1995, Brewster and Broughton 1999, Fricke and Ahearn 2002, Hansen 2002, Shapiro 2002, Katz 2004, Chang 2005, Katz 2006, Katz 2012, Smith 2013, Hansen 2015, and the DVD *Scratch* (2001).

TWO

Page R and the Art of the Loop

The Fairlight CMI Series II, IIx, and III

With the sample time on the Fairlight CMI Series I limited to one second, users employed it for inserting short sounds into recordings. This changed with the release of the CMI Series II in 1982, the Series IIx in 1983, and the Series III in 1986.[1] These contained a built-in sequencer called Page R (or Real Time Composer), which enabled users to build rhythmic patterns of sampled sounds. J.J. Jeczalik and other members of Trevor Horn's production team started to use the Fairlight CMI with other digital technologies to add loops with samples to recordings by Malcolm McLaren and the Art of Noise. This mirrored the hip-hop aesthetic of isolating and repeating rare breakbeats with analogue technologies such as turntables and vinyl, or magnetic tape. Using archival research and material from an interview with Jeczalik, I look at the use of the CMI Series II, IIx, and III between 1983 and 1988. I focus primarily on three things: (1) its use to construct collage-like recordings, which were inspired by the cut and scratch turntable techniques of hip-hop DJs; (2) the use of Page R and other sequencing technologies to create musical performances that were 'strictly in time'; and (3) its use to sample pre-existing sound recordings and manipulate them in new ways. In this chapter, I continue to trace the history of Fairlight Instruments until the company closed in 1988 and follow the instruments, users, and non-users that helped to shape the sounds and practices of a loop-based aesthetic in pop/ular music.

Following the Users and Non-Users: Trevor Horn and the art of delegation

The focus on users in Science and Technology Studies (STS) developed as a counterbalance to the privileging of actors like scientists, designers, and engineers in the shaping of new technologies. Users, though, are not a homogenous group. Sally Wyatt draws attention to different categories – for example, former users as well as current users. We might also extend this to first-time users versus experienced users. There are non-users who 'resist' or 'reject' a technology for reasons that do not fit the traditional narrative of access being restricted due to socio-economic circumstances. Wyatt asks: 'What exactly does it mean to be a user? How is it defined? Is it possible to distinguish between non-users and non-owners?' (2003, p. 76). The first chapter featured a range of users: the designer of the Fairlight CMI (Peter Vogel) who demonstrated it to first-time users, users who were owners of a CMI (Richard Burgess, Peter Gabriel, Stevie Wonder, and Herbie Hancock), and users who were non-owners (Kate Bush). In this chapter, I begin with a well-known music producer and owner of a Fairlight CMI who never learned to become a user, for reasons relating to time rather than money.

Trevor Horn became associated with the Fairlight CMI while producing records in the 1980s for acts like ABC, Malcolm McLaren, Yes, and Frankie Goes to Hollywood. He bought a Fairlight when the owner of the one he had access to joined another band:

> Geoff [Downes, Horn's partner in The Buggles] had a Fairlight but he'd gone off to form Asia. So, when he went I bought a Fairlight. That, actually, I must admit, freaked my wife out because it was £18,000 and that was a fortune back then! There were only four of them in the country and I had one of them. But what was even more important was I knew what it was capable of, because I understood what it did. Most other people didn't understand at the time – sampling was like a mystical world.
>
> (quoted in Peel 2005, p. 52)

Horn had previously admitted to possessing very little technical expertise in the recording studio: 'You ask anyone I work with, I never touch anything. I've got no idea how to work a Fairlight' (quoted in Hoskyns 1984, p. 26). He had also been dismissive about the lasting impact of the CMI and other new technologies: 'All the equipment, the Fairlights and so on, are just another passing fad. I'm beginning to hate all of that stuff. . .' (*ibid.*). Rather than confusing the owner with the user, I want to pay closer attention to the collective processes of music making and follow another social actor, J.J. Jeczalik, who was the one credited with programming the Fairlight CMI on recordings produced by Trevor Horn.

(i) J.J. Jeczalik: Sessions, Recordings, and the Life of a Freelance Fairlight User

A geography graduate from Durham University with little formal musical training, Jeczalik taught himself how to use the Fairlight CMI. His musical career began when he was employed as a roadie for the band Landscape. Their drummer, Richard Burgess, advised Jeczalik to 'get into computers and learn how to type' (Jeczalik 2011) rather than learn how to play drums. As a session musician, Burgess also played drums on recordings by The Buggles, the duo consisting of Horn and Downes. With Jeczalik working as a roadie and part of this network of musicians, Horn and Downes made him an offer of employment: he became their keyboard technician. This included using the Fairlight CMI, which left Jeczalik overwhelmed when it was first demonstrated to him:

> That was with Trevor and with Geoff. I think we went to the store that was selling them and they explained what it could do. I was just completely blown away. I didn't sleep for about a week because I just thought it was incredible. It was just an amazing thing. They gave me a quick demo. Plug this in, do that, turn that, hit that, do this, and I can play back my voice. I could suddenly see it all. In this blinding flash, I thought 'Blimey, this is incredible'.
>
> (*ibid.*)

Rather than using it to replicate the sounds of acoustic instruments, Downes planned to use the Fairlight CMI as a single replacement for the large number of keyboards that he used during live performances. Jeczalik began by asking himself:

> 'Why don't I start off by sampling all his keyboards?' but it became very apparent that the quality wasn't good enough. Because of the polyphony issues – it was only 8-note polyphony in those days – we could never get enough sounds on it from a live context to have one keyboard. I think we both envisaged that we could use the one Fairlight keyboard to do everything but it rapidly became clear that that wasn't going to be the case. It was too slow. There wasn't enough polyphony and it wasn't designed for that sort of work but we used it to do a lot of sampling in the studio. Just editing sounds, trimming them, cutting them.
>
> (*ibid.*)

His initial excitement about using the Fairlight CMI to sample external sounds was tempered by the technological limitations and fidelity issues faced by users. Language associated with analogue technology – the splicing of magnetic tape – is employed to explain how the CMI was used instead for digitally editing the sounds of recordings.[2]

Jeczalik described his early attempts to use the Fairlight CMI as being as much about learning to use a computer as it was learning to use a musical instrument. Unlike the more studied approach of a user like Burgess, he chose to ignore the instruction manual:

> In the very early days when I was working with Geoff Downes on his Fairlight I didn't really understand the process. I had a general understanding of what was going on. You put a sound into the computer, some interesting stuff happened, then you saw some lines on the screen and you pressed a key and it came back. When I first started we were just playing with it right from the get go and didn't approach it from a technical point of view. We just sort of plugged it all in and played with it because the initial manual that came with the Fairlight was about twenty pages long.
>
> (*ibid.*)

Approaching the CMI without the embodied knowledge of a pianist or keyboard player, Jeczalik's lack of musical training and technical experience may have been an advantage as he experimented with it.[3] For him, the appeal of the technology was that formal training did not seem to be a pre-requisite for using it in creative ways. Jeczalik also saw an opportunity to position himself in the field as a user with expertise:

> I think the potential was that you could do pretty much anything you wanted and it was very early in the game. I could see that I could build a niche for myself in terms of doing something that was creative. I didn't play as such. I'm a one-fingered keyboard player so it gave me the opportunity to think, well actually, I could really do some interesting stuff with this thing. I didn't know what specifically at the time.
>
> (*ibid.*)

One of the recurring themes in the discourse about digital sampling has been that it offers everyone the opportunity to become a musician. In a critique of articles by rock journalists who celebrated sampling as a subversive practice, Simon Reynolds and David Stubbs wrote that '[s]ampling has been championed as a new punk – both a repossession of control from the industry, and a liberation from the inhibiting effects of notions of expertise' (1990, p. 168). As the role of some session musicians morphed into that of session programmers in the early 1980s, Jeczalik was recognised precisely for the expertise he had developed as a Fairlight CMI programmer. He was hired for his knowledge about how to operate a musical instrument that only a small number of people had access to.[4]

Before going on to look at how Jeczalik and others used the Fairlight CMI Series II/IIx after the introduction of the sequencing software, Page R, I want to briefly sketch out how he used it on recording sessions before 1983. Jeczalik received more employment opportunities as a result of the concerns expressed by the Musicians' Union and fears about how digital synthesizer/sampling instruments were being used:

> There was a lot of press at the time, which did me no harm in terms of getting work. People were going: 'It's the end of the orchestra. This is going to take over everything. Musicians are going to be redundant'. And a lot of people wanted to see what all the fuss was about. So, I was going on sessions with this kit, sampling things, and explaining that actually it had a very short sample time. To loop it you had to have all the tuning aspects and everything going for you, otherwise it sounded pretty bad to be honest.
>
> (Jeczalik 2011)

For this user, the ability to imitate the sounds of acoustic instruments with the Fairlight CMI was exaggerated. The producer and recording industry veteran, George Martin, was also unimpressed:

> One of the sessions I was working on was with Paul McCartney and he'd had a trombonist in. This was with George Martin who said, 'Let's put the note in from the trombone and then we can have a horn section'. So, sure enough, I put it in and tried to loop it. It was really difficult to loop it to get any sustain. He pressed the call and he turned to me and said, 'That doesn't sound much like a horn section does it?' And I went, 'No. Well, it's not. That's not what it is. It's a sample of trombone played with four notes'. And it became very apparent at that moment that it was pointless sampling other instruments.
>
> (*ibid.*)

It is interesting to speculate whether Martin, with experiences of cutting and splicing magnetic tape while working with The Beatles at Abbey Road Studios in the mid-to-late 1960s, recognised the creative possibilities of using sampling instruments or their application as time and energy-saving technologies. However, Jeczalik was unable to discuss it with him. One of the other consequences of this session was that it forced Jeczalik into making a decision about becoming a fully self-employed programmer.[5]

As a freelance user/programmer of the Fairlight CMI, Jeczalik worked on a number of recording sessions including those for Kate Bush's *The Dreaming* (1982) before she became a CMI owner *and* user. Like Burgess, Jeczalik has fond memories of working in the recording studio with Kate Bush and describes a session he was part of in 1981:

> The main thing we did with her was a thing called 'Get Out of My House' and we went around sampling doors closing in Townhouse Studios for a day, which was quite a lot of fun actually because you started to hear how all the doors sounded. Obviously doors just close and you don't think about it but after a while we started going around thinking, 'That's an interesting door'.
>
> (*ibid.*)

Jeczalik first collaborated with Horn on tracks that were released on Dollar's *The Dollar Album* (1982) and ABC's album, *The Lexicon of Love* (1982). On the latter, Horn was the producer and Jeczalik was credited with Fairlight programming. The sounds of the CMI do not dominate the sound of the recordings. Jeczalik explained that '[t]he role for the Fairlight at that time was just popping in some interesting bits and pieces here and there' (*ibid.*). This can be heard on two examples of the 'New Pop' ABC had begun to explore along with other post-punk groups like Scritti Politti.[6] At the beginning of 'Date Stamp', the breathy sound of a synthesizer is interrupted by the ringing sound of an old cash register being opened, providing a motif to accompany the song's lyrics about the supply and demands of love. Jeczalik told me: 'I can't remember where I got the cash till from but it's a fantastic attack and sustained sound. It's very distinctive' (*ibid.*). Before the CMI could be connected to other music technologies like the Linn Drum and before Page R was available, Jeczalik was inserting the same sounds of everyday life into recordings that had been reproduced in progressive rock using analogue technologies.[7] The other example from *the Lexicon of Love* where the use of the Fairlight CMI is audible is the track '4 Ever 2 Gether', which begins with the sinister tone of a synthesized voice repeating the word 'evil'. Jeczalik explained:

> I recorded someone – it was Julian, one of the engineers, I think – saying 'Speak no evil'. We played that into the track and actually that was the first time I had played on a record. It was my first overdub. It went, 'Speak no evil' and then I just de-tuned the 'evil evil evil' down into a really low menacing sound.
>
> (*ibid.*)

One of the reasons Jeczalik felt excited when seeing a Fairlight CMI for the first time was because voices could be sampled and this was the first thing he did in the studio. It is an early example of the ways sampling technologies were used to record and play back the distorted sounds of human voices. Producers at this time were also using E-mu's Emulator keyboard and technologies such as the AMS DMX 15-80 digital delay line to re-create the effects of stuttering and experiment with other forms of vocal manipulation.[8]

(ii) 'Sampling' the Sounds of the World: Collages, Loops, and Copyright

On Malcolm McLaren's album, *Duck Rock* (1983), Jeczalik began constructing collages and loops from studio recordings and, in this case, studio recordings made 'in the field'. His use of the Fairlight CMI shows how a wider set of musical practices associated with sampling technologies were developing. After the implosion of the Sex Pistols, McLaren had moved from managing bands to making records. He recruited Horn as his producer and Gary Langan as his engineer on a proposed trip around the world to record folk dances, although the only stop offs were in Soweto and New York. The trio had the Fairlight CMI available to them but the sounds that provided the basic materials for *Duck Rock* were recorded and stored on magnetic tape. Horn stated that he, Langan, and McLaren

> . . .had two options. Either we could take a Fairlight. . .
> copy the rhythms from all the different sources Malcolm
> had and then go out and make songs from that, or we
> could actually go out and get the sounds from the actual
> people, capture the real things on a Nagra [a two-track
> tape recorder].
>
> (quoted in Bromberg 1989, p. 260)

With one second of sample time available, they had little choice but to opt for a more portable recording device. They employed musicians from a variety of continents – living in the towns and cities of North America and the townships of South Africa – to perform on the recordings. This led to a legal dispute over copyright that predated the controversy about the inclusion of performances by South African musicians on Paul Simon's *Graceland* (1986) (Meintjes 1990) and debates about the ethics of sampling recordings featuring non-Western musicians (Feld 2000; Taylor 2003; Théberge 2003).[9]

After the sounds of performances were collected using magnetic tape, Jeczalik employed the Fairlight CMI on the sessions for *Duck Rock* to insert a range of pre-recorded sounds into new recordings. These took the form of collages and loops as well as what others involved in the process referred to as adding 'bits and bobs'. Anne Dudley, a classically trained musician who provided string arrangements on the ABC album, also missed out on the field trips to Soweto and New York: 'I wasn't involved in the process that they went through with *Duck Rock*, going around the world collecting various bits and bobs. I was only involved when they started putting them all together, trying to collate it into some sort of sense' (quoted in Buskin 1995, p. 108). Jeczalik described his part in the process of locating the pre-recorded materials, or what he calls the 'bits and pieces':

> Gary [Langan] and Trevor [Horn] went off to South Africa and came back with tapes and tapes of recordings. I used to go into the back room and pick out sounds that I thought sounded interesting. I put them into the Fairlight and then we just used to play around with them to see what would work. It was very, very experimental. We were just experimenting with bits and pieces that they'd picked up. Gary would put some tapes together, half-inch tapes of things that he thought were cool and interesting. I'd sit in the back room and bung them in the Fairlight for a couple of days and then we'd go into the studio.
>
> (Jeczalik 2011)

Along with the Fairlight CMI, hip-hop DJs used turntables and cut and scratch techniques to insert pre-recorded sounds into songs like 'Buffalo Gals'. The sleeve notes to the single contained instructions about how to make a similar record and what instruments to use, continuing punk's DIY ethic of sharing knowledge and demystifying the production process. Expensive digital synthesizer/sampling technologies were not included as a requirement.[10] However, the 'rhythm boxes' used in the making of 'Buffalo Gals' were drum machines containing digital samples of acoustic drum sounds. Horn explained:

> By the time I did the McLaren record I'd bought an Oberheim sequencer and drum machine, a DSX and a DMX. I told the World's Famous Supreme Team to tell me their favourite drum beat. It took a couple of hours for them to actually communicate it to me, but once I'd got it, that was 'Buffalo Gals': 'du du – cha – du du – cha'. That was done on this DMX and DSX and they just scratched on top of that.
>
> (quoted in Peel 2005, p. 53)

As it was not possible yet to use the Fairlight CMI to reproduce drum patterns from pre-existing sound recordings due to the lack of available memory, Horn created rhythms using drum machines such as the Oberheim DMX (1981). Like the Linn LM-1 Drum Computer, it contained samples of what the designers called 'real' drum sounds.[11]

Before the introduction of Page R on the Series II, IIx, and III, Jeczalik was using the Fairlight CMI to record and play back what he called 'short punchy sounds' (Jeczalik 2011). On *Duck Rock*, he was also starting to loop sounds from the sample library and this can be heard at the end of the track 'Punk it Up' as it draws to a close.[12] When I played this to Jeczalik over the telephone, he recognised its source immediately:

> That's ORCH5 in there from the Fairlight, which was one of the library sounds. I think that was just a loop that I put together because I was goofing around at the time. Trevor said, 'What have you got?' and I went, 'I've got this' and we just sort of looped it. I don't really remember to be honest. I'm not being vague. There was a voice sample as well but quite short and obviously that was the Supreme Team doing the voice over. That was live. That was the sort of thing we were doing, coming to the end of a track, you need something in here, something different, and I just spun out some sounds to see what stuck.
>
> (*ibid.*)[13]

This is an early example of the Fairlight CMI being used to create 'loops' of digitally recorded sound rather than adding digitally recorded sounds onto loops of magnetic tape. However, while the pre-recorded sample library was the source of the sounds – the ORCH2 rather than the commonly misidentified ORCH5 sound – Jeczalik had to employ another device alongside the Fairlight CMI to create the loop. He explained:

> I think that ended up in something like an AMS digital sampler. It was looped in that. That would be my guess. I don't remember whether we had the sequencer or not. I think the sequencer was around about then or just after but I have a suspicion that because Gary [Langan] and I used to work quite a lot together, he would use the AMS digital delay and sample a mono block into that and just repeat it.
>
> (*ibid.*)

The AMS device Jeczalik refers to was not a digital sampler. It was an AMS DMX 15-80 digital delay line that could be used to trigger loops in a similar way to the contemporary practice of using digital delay pedals to repeat parts of performances in real time.[14]

Jeczalik's stimulus for using the Fairlight CMI and other technologies to loop sounds came from the historical uses of tape loops in popular music and the emerging practices of hip-hop. It was also fuelled by a discourse about creativity that included ignoring conventions, not following instructions, and a desire to do things differently:

> It was inspired by all of those things. It was inspired by Malcolm who just used to say, 'Well, why not?', which was an interesting point of view because people will go, 'You just don't do that' and he would go, 'Well, why not?' and you'd go, 'Okay, let's do it'. He was challenging the concept of what a record was and I suppose the looping came out of all of that, of the guy scratching. Tape loops had been used for many years. Then there was looping with samplers but we didn't use the Fairlight for looping at that time. I'm fairly confident we didn't use it as a sequencer but it would have contributed to something that would have ended up probably in a digital loop, in a delay or something.
>
> (*ibid.*)

Anne Dudley also paid tribute to McLaren's preference for breaking rules: 'He was outrageous – he showed us that anything is possible' (quoted in Husband 1985, p. 20). She also explained that the production techniques on *Duck Rock* were responsible for stimulating musical ideas that were developed further during their work with the Art of Noise. With Page R, the Fairlight CMI could now be used to both sample sounds and arrange them rhythmically using an in-built sequencer. Horn describes how:

> It was an amazing time because it was all exploding. Just as the McLaren thing came to an end, Page R arrived on the Fairlight. And that was gobsmacking because that was the first time you heard those sorts of sounds sequenced. And that's where the Art of Noise came from.
>
> (quoted in Peel 2005, p. 53)

Before looking at Jeczalik's approach to digital sampling with the other members of Art of Noise, I want to examine how he and Horn used the Fairlight CMI Series II/x along with other technologies to try and create recordings that were 'in time' and fully quantised.

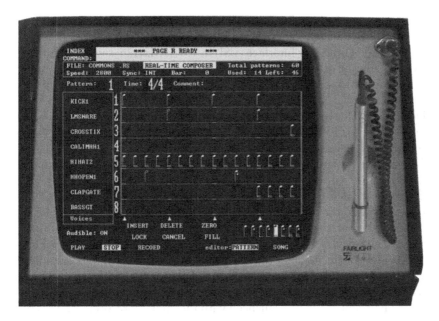

Figure 2.1 Page R on the Fairlight CMI Series II

Following the Fairlight CMI Series II and IIx: The Real-Time Composer (Page R)

The Fairlight CMI Series II and the Series IIx both contained three compositional programmes: a real-time multitrack sequencer (Page 9), a non-real time Music Composition Language (Page C), and a Real-Time Composer (Page R). An advertisement for the CMI Series II in the US magazine *Musician* explained how:

> The Real-Time Composer is our most recent development and continues to cause considerable excitement among CMI users. This high-speed function allows rapid development of complex phrases, making it particularly suitable for rhythmic compositions. All pitch, timing, and dynamic information is recorded and displayed while an automatic quantising facility corrects playing inaccuracies. Editing may be performed live or through the typewriter keyboard.
>
> (Fairlight 1983b)

Enabling users of the Fairlight CMI to build rhythmic patterns using a number of different instruments, Page R has been called 'the first ever graphical pattern-based sequencer' (Leete 1999, p. 255). Synthesizers with MIDI sequencing became available after MIDI was agreed upon as a Universal Synthesizer Interface (USI) in 1983. Machines were beginning to speak to each other digitally.[15] As well as being able to sequence sounds using Page R on the Fairlight CMI Series II and IIx, another significant development was the design of an interface board called a Conductor. Designed by Steve Rance, a Fairlight CMI user in London, who went on to work for Fairlight Instruments in Sydney and Fairlight US in New Jersey, they were sold by Syco Systems and cost £800: 'I used to hang around Syco Systems all the time. Having access to all of the 'other' equipment that at that time could not be synchronised easily, pushed me into designing something to make them all talk to each other' (Rance 2015). Users now had the option of connecting the Fairlight CMI IIx with other technologies such as drum machines. Horn explained how '[a]t that time there was a device called a Conductor, which enabled you to synchronise a Linn drum machine with a Fairlight, and to us it was the most incredible thing ever' (quoted in Buskin 1994, p. 40). Rather than Horn recording 'real' drummers playing in real time, digitally sampled sounds on the Fairlight could be sequenced with the digitally recorded sounds of 'real' drums that were stored in the memories of drum machines.

(i) Digital Sampling, Drum Machines/Computers, and Sounding like Machines

The use of Page R and the Conductor to connect together sounds that had been sampled with the Fairlight CMI with sounds from drum machines was integral to the sound of 'Relax' by Frankie Goes to Hollywood (1984), which was produced by Horn. The kick drum sound was created by combining a bass sound from the LinnDrum and an E note from a bass guitar sampled into the Fairlight CMI.[16] He described how:

> It was a combination of Page R and the Conductor and locking it to a Linn drum machine. So, the basic track was eights [eighth notes] running in a Fairlight, fours [quarter notes] on a bass, and a set of Linn drum machine patterns locked to Page R played on top of each other. It was an amazing feel.
> (quoted in Peel 2005, p. 53)

Jeczalik's account of what was going on in the studio during the making of 'Relax' explains why sequencing and sampling were important to the rhythmic sound of the recording. It is worth quoting at length as it shows how the final version approved by Horn was the result of neither accident nor design but the principle of 'goofing around':

By then we'd got the sequencer on it, Page R, which was an eight-track monophonic sequencer. You could put eight notes on at one time. What happened was we tried recording this track with a band and with Ian Dury's backing band and spent months on it. I was working with Andy Richards, keyboard player, Steve Lipson, who was engineering at the time, and Trevor [Horn] at Sarm Studios. Trevor went home and we were kind of getting nowhere really. It just didn't feel right. I remember saying: 'Let's just put an eight-bar loop together'. I had a piano sample and put that in doing eighth notes, 'dum dum dum dum dum dum dum dum'. Someone programmed up the drum box to do a 'boom tack boom tack boom tack' type thing and then Steve Lipson got his guitar out. Andy was playing keyboards and I had a whole load of samples from the band doing backing vocals. We just started goofing around, literally goofing around, and cranked the volume up. We were really enjoying ourselves and started to realise that we actually had something. I was putting in piano eighths and some samples of the backing vocals that come with little fills here and there. The one that goes [imitates wah wah guitar sound], that's one of mine as well. So, we just built this thing up and we had this hell of a racket going. Trevor came back and said, 'What the bloody hell is this?' No, he didn't swear. He said, 'What on earth's going on?' or something. We turned the volume down and I think it would be fair to say we all looked round a bit sheepish and thought we'd done something wrong. We went, 'It's just a loop'. He said, 'No. No. No. It's brilliant' [Laughs].

(Jeczalik 2011)

While claiming to be a non-user of the Fairlight CMI and other studio technologies, Horn returned to programme the LinnDrum after being absent from the recording process. Using Page R to sequence sounds that had been sampled using the Fairlight CMI or using a device like the Conductor to connect the Fairlight CMI with Linn drum machines, Horn and his team could create rhythms that they considered to be 'strictly in time'.[17]

Before the availability of drum machines, sampling instruments, and the ability to connect them together, Horn used a variety of methods in the recording studio to elicit performances from drummers that were as metronomic as possible. This often led to exhaustion for session musicians taking part in these tests of endurance. Horn stated:

> I do remember that by the time we'd finished playing 'The Plastic Age' [by The Buggles], Richard Burgess was pale! He was so worn out because we insisted that it sound perfect and that he played it perfectly. And the funny thing is that when you listen to it, it sounds like a drum machine. Both tracks sound like drum machines because at the time we were so manic about them having that spot-on perfect techno feel, not some sort of bullshit Elton John groovy-album feel.
>
> (quoted in Peel 2005, p. 51)

On 'Video Killed the Radio Star' (1979), the first single from the album, *The Age of Plastic* (1980), the drums were produced differently to achieve the same effect. It features individual drums sounds that were recorded and re-constructed using the mixing desk to sound as if they were programmed into a drum machine. Horn explained:

> I got Paul Robinson to play his kit one drum at a time – the snare, bass drum, and the hi-hat – and I recorded them on separate tracks, then used the sounds like a drum machine, punching him in and out on the desk. Paul said, 'That sounds fucking awful, just like a machine'. I said, 'Great, that's exactly how I want it to sound'.
>
> (quoted in Cunningham 1998, p. 271)

While representatives of organisations like the Musicians' Union were concerned at this time that the use of technologies like synthesizer/sampling instruments and drum machines were creating fewer opportunities for performers, the ability of humans to imitate machines that could be used to create metronomic rhythms was becoming easier.

(ii) The Art of the Loop and the 'Recycling' of Recordings

Horn cited the introduction of Page R on the Fairlight CMI and the sequencing of sampled sounds as one of the reasons for the formation of Art of Noise. Along with the collaborations on *Duck Rock*, the project was also shaped by Horn's production of Yes's album, *90125* (1983). On tracks such as 'Owner of a Lonely Heart', Jeczalik programmed sounds using the Fairlight CMI. His use of samples was different to 'the sounds of everyday life' that groups like Yes and Pink Floyd had incorporated into their recordings using analogue technologies in the 1970s. On *90125*, Jeczalik used the piano keyboards of the Fairlight CMI to organise short samples into melodic patterns or to add rhythmic effects:

I used to go in and out doing bits and pieces. I sampled
some of the backing vocals. There's some 'dums' in there
that Chris Squire sings. I sampled Alan White's bass
drum snare. I sampled some of him doing drum fills. In
terms of 'Owner of a Lonely Heart' itself, Trevor put
a cassette on and said, 'I want to sample these drums',
which I did. We kept listening and the stabs that came
along, knowing by then what the Fairlight sounded like
and what it would do to them, I said, 'I think those are
really cool'. That's where I got those sounds.

(Jeczalik 2011)

The looped drum patterns on Art of Noise's recordings were formed,
then, by sampling the drum fills of musicians playing acoustic drum kits.
Engineer Langan described how this was the consequence of overwork
and overtime during the sessions with Yes:

After about seven months of working virtually every day
of every week at a variety of studios, I was beginning
to see green men climbing the walls. We had been up at
AIR in Oxford Circus to cut a track but it was scrapped.
I kept the multitrack though because the drum sound
on 'Leave it' was just phenomenal. A month later, when
the band had gone home one night, J.J. and I had the
idea for putting the drums from this multitrack into the
Fairlight as a complete sample. The idea wasn't to have
separate samples of the bass drum, snare, and hi-hat, like
everyone was beginning to do with AMSs, but have it as
a composite of the whole kit. So that's where the drum
sound on 'Close (to the Edit)' came from. J.J. and I effec-
tively recorded the first Art of Noise single that night,
although to us it was a demo. We just looped the drum
sample and added a few other things on top.

(quoted in Cunningham 1998, pp. 309–310)

It is unclear how this was possible with only one second of sample time
on the Fairlight Series II/IIx but Jeczalik confirms Langan's version of
recording studio events. Rather than a light bulb moment or the product
of skill and expertise, he also describes the drum loop on 'Close (to the
Edit)' as the result of a mistake caused by a lack of concentration:

It evolved basically because we used a lot of the sounds
from the Yes sessions, notably the drum sound, which
became the bedrock of what The Art of Noise was all

about. By then I'd been working on hundreds of sessions and had about a hundred discs of sounds. So, one day we'd finished a session and Gary had an idea to stick around. We stayed and he got the drum sound from a Yes session. We put it in the Fairlight and basically off we went. It evolved again because basically it went in as a loop. What happened was I wasn't paying attention and I sampled it on the snare beat so rather than going sample 1-2-3-4 in the bar, I went sample 2-3-4. We used to call the sample 'tack boom boom' because that was the sound it made. It was snare drum, bass drum, bass drum. Interestingly, when we started looping it, because it worked as a loop, it still made a bar of four or whatever. So, when the loop started happening it just had this incredible feel. It was complete luck or misjudgement on my part and so that became the backing track for 'Close (to the Edit)'.

(Jeczalik 2011)

At this point, Jeczalik and Langan were sampling sounds from recordings they had been working on. Armed with a theoretical scaffolding about raiding the sounds of the twentieth century, Art of Noise began to sample sounds from pre-existing recordings as well as more 'natural' sounds, that is, 'the sounds of everyday life' and its modes of transport.

As well as Jeczalik and Langan's use of the Fairlight CMI to construct loops from pre-existing studio recordings on 'Close (to the Edit), they also sampled the sound of a car starting at the beginning of the track. This was also used on Frankie Goes to Hollywood's first album, *Welcome to the Pleasuredome* (1984) and Jeczalik is happy to admit this was part of a process of 'recycling' rather than trying to find new or original sounds:

We were always recycling and chopping up. For example, on 'Paranoimia' by Art of Noise, we had some kids in from a local drama school and we had them say things like 'The Art of Noise are paranoid'. I took the sound 'paranoid' and then flipped the middle bit, got a section, and reversed it, and it came out as paranoimia. I created a word and that became Paranoimia. Para-, Para-, Para-, Paranoimia. That was where we were going in terms of recycling stuff. We'd chop a bit out, reverse it. You know, for example, the car starting. It's backwards and it's all over 'Ferry Over the Mersey' by Frankie Goes to Hollywood.

(*ibid.*)

Jeczalik was using his library of samples as a palette of sounds and distinctive samples appeared on a number of tracks. For example, the loop from 'Close (to the Edit)' was chopped up and the individual sounds added to other recordings such as 'Beat Box'. He explained how the 'recycling' process created incongruity between the recorded sounds of acoustic instruments and sounds that had been reproduced using the Fairlight CMI:

> It was all just serendipity really. Gary [Langan] understood very well how the Fairlight operated and what it was good at. He had an incredible way of making the drum sounds sound much huger than they actually were. When Anne [Dudley] came in and started putting keyboards on which were real – we often used real pianos and so on – the contrast and the sonic quality was just extraordinary because you had a very low bandwidth drum track thrashing away in the background and then these real sounds played over the top that had incredible high-definition. It created an incredible soundscape, which I didn't really appreciate at the time. I understand now what happened and why the keyboard sounded so good. It was because the Fairlight sounded so awful [Laughs].
>
> (*ibid.*)

Kate Bush and members of the BBC's Radiophonic Workshop used the Fairlight CMI to reproduce the 'real' or 'natural' sounds of everyday life. For Jeczalik, 'real sounds' are those of acoustic instruments like the piano. He distinguishes between these and the sounds reproduced by the Fairlight CMI, which had lower levels of sound quality.[18]

(iii) Raiding the Twentieth Century: The Sounds of the Futurists and the Art of Noises

One of the theoretical ideas constructed for Art of Noise by journalist Paul Morley was the concept of 'raiding the twentieth century'.[19] His idea was shared by other members of the group who wanted to create collages of high and low culture. In an interview with *No. 1* magazine in February 1985, Dudley described how '[e]verything is available to us. We're influenced by anything. J.J. has a passion for Mahler. I have a passion for Stravinsky and Holst. And Nat King Cole' (quoted in Husband 1985, p. 20). However, for Jeczalik, there was a tension between ideas relating to modernism (noise as music) and post-modernism (the collapse of high and low), and the practical activities of music making:

It always started with the sound. It always started with
music and it would be safe to say that I didn't consider The
Futurist Manifesto and all that side of it at all when creating the music. I just wanted to create stuff that sounded
good and exciting and interesting and challenging and taking the kind of Malcolm McLaren mould of going, 'Why
not?' There were fifty different elements in some of the Art
of Noise tracks. We were just getting stuff to sound interesting and exciting to us. My view was if you provoke a
reaction then there will be an awful lot of people who will
love it and it's got to be exciting. Although, on the other
hand, we made 'Moments in Love' as boring as we could.
It became a de facto love song and it's still in the charts in
America thirty years later.

(Jeczalik 2011)

At more than ten minutes long, 'Moments in Love' was part of Morley's
objective of 're-defining what a pop group is' (quoted in Martin 1984,
p. 35) and what they were allowed to do within the confines of sales
chart rules. Continuing the theme of recycling, it is also interesting for its
use of sounds from a previous Art of Noise recording, 'The Army Now'.
According to Jeczalik, one of the group's aims was to 'juxtapose odd and
wondrous things in different ways' (quoted in Mico 1985, p. 15). This
extended to his use of the Fairlight CMI to sample the sounds of pre-
existing recordings by other artists as well.

'The Army Now' from the 'Into Battle' EP (1983) contains elements that
were used on 'Moments in Love' and 'Close (to the Edit)'. It lasts approximately two minutes and contains a sampled phrase of the three words in
the title. The phrase is repeated, as are the individual words 'army' and
'now', which were manipulated in different ways using the Fairlight CMI.
These are sampled from a pre-existing recording but Jeczalik was reticent to
talk about it or explain their source due to the fear of legal action: 'To this
day, I'm still nervous about all that. I'll probably die being nervous about
it' (Jeczalik 2011). To disguise the sounds from pre-existing recordings and
to avoid detection for unauthorised copying, Art of Noise used different
recording studio techniques to make it difficult to identify the source of
their samples. It was also to create more 'interesting' sounds:

There was no clearing of samples. It was so new and I
was putting things backwards and sideways and putting them in reverse echo. Even now probably I could
listen to some of that stuff and I couldn't tell you where
it came from because we disguised it so well. Gary
[Langan] used to do weird and wonderful things to the
sound so that you couldn't tell where it came from.

> It wasn't necessarily to disguise it. It was just to make
> it interesting. It had to stand or fall on what it sounded
> like at that moment. Nothing else.
>
> (*ibid.*)

Morley had proposed a raid on the sound recordings of the twentieth century and it was underway. However, it was restricted to small fragments of pre-existing recordings because of issues over the quality of sounds that could be reproduced using the Fairlight CMI Series II/x, the maximum sample time of only one second, and the threat of legal action as a result of infringing copyright law.

While the marketing of synthesizer/sampling technologies encouraged users to create the sounds of the future, they could be used to re-create as well as sample the sounds of the past. Jeczalik used the CMI Series II to create sounds that were more like those produced with guitar amplifiers in the 1960s. In an interview in 1993, he explained how:

> The interesting thing about the Fairlight Series I and
> II is that your samples come back radically different.
> They sound as if you've put them through a 100-watt
> Marshall amp. For me that adds an element of rock 'n'
> roll, which I've always valued and exploited.
>
> (quoted in Tingen 1993, p. 52)

The low sample rate and bit rate caused problems for Jeczalik and he describes buying a Fairlight CMI Series III because 'the quality [of the Series II] was doing my brain in' (Jeczalik 2011). However, Jeczalik also admitted to deliberately using a lower sample rate of 15 KHz instead of 44.1 KHz when programming the Series III to make it sound like the Series II. This was because of the 'grunginess it gives you' (quoted in Tingen 1993, p. 52) and to 'make things sound dirty and distorted, and rock 'n' roll' (quoted in Tingen 1996b, p. 98). Jeczalik also appreciated the unpredictability of the Fairlight CMI compared to the samplers that companies like Akai introduced in the mid-1980s:

> I bought an Akai sampler, which I never really got on
> with but they had more time on them. I always liked the
> sound of the Fairlight and that was part of what I did.
> It was part of turning up and plugging this thing in and
> sampling it. It was like a giant guitar effects pedal. You'd
> put your thing in but you weren't really sure what would
> come out. When the Synclavier and the Fairlight Series III
> came out, they didn't interest me because they sounded
> too good. There was no modification going on there and
> I always liked the slight mystery. You put something in

and it would sound fantastic. You put something else that you thought was going to work really well and it didn't work at all. There was a bit of a dark art and mystery to it all whereas if you have a high-quality sampler that just throws back what you've got then you have to start work on making it different.

(Jeczalik 2011)

Users of the Fairlight CMI did not necessarily want the same control over sounds that its designers aimed to provide: Jeczalik was keen to avoid digital 'perfection' and predictability. The designers of synthesizer/ sampling technologies at Fairlight Instruments and New England Digital strived to continuously improve the quality of the sounds that could be digitally recorded and reproduced by their instruments. However, a user like Jeczalik preferred the older model of the Fairlight CMI to the newly released one precisely because of the lower levels of sound fidelity and the timbres they produced.

The Fairlight CMI Series III and the Commercial Failure of Fairlight Instruments

With up to 14Mb of RAM, Fairlight Instruments launched the CMI Series III in 1986. They replaced the monitor that came with the Series I and II with a 12-inch Video Display Unit (VDU) and the software moved closer to a WIMPS (Windows, Icons, Mouse, Pointers, Systems) interface. Instead

Figure 2.2 Peter Vogel (left) and Kim Ryrie with the Fairlight CMI Series III

of a light-pen, waveforms were edited with a graphics tablet and stylus that was attached to an alphanumeric keyboard. With one six-octave keyboard and one 8-inch floppy disc drive rather than two, the Series III used 16-bit digital-to-analogue converters and contained twelve microprocessors running Motorola's OS-9 operating system. The basic cost was £25,950.[20] A review in *Sound on Sound* magazine described the Series III as 'a unique instrument that is sure to have a long-term place in the development of music technology' (Elen 1986, p. 55). However, by September 1987 Syco Systems had stopped acting as the sole UK distributor of the Fairlight CMI because of slow sales and Fairlight Instruments went into receivership in the last few months of 1988 (Tingen 1996a). When the Series III was launched, Kim Ryrie expected around half of its users to employ them for producing film soundtracks.[21] After the demise of Fairlight Instruments, a new company called Fairlight ESP (Electric Sound and Picture) was started in April 1989 to focus solely on the post-production industry.[22]

The successful invention of electronic musical instruments is judged by the commercial availability of these technologies and their subsequent adoption by musicians and users. The failure rate, though, is high. Paul Théberge speculates on the reasons for this:

> In some cases, the failure of these instruments may have been due to a simple lack of business acumen on the part of their inventors. Inventors seldom possess the business skills required to manufacture and market a musical instrument successfully, even one superbly designed.
>
> (1997, p. 41)

When they started Fairlight Instruments, Vogel and Ryrie had few business skills. They did not understand the music industries but were initially successful in connecting the worlds of design and use. In 1980 they won an award for the Qasar Dual Processor Microcomputer System from the Industrial Design Council of Australia. Two years later, the company was lauded for the successful export of its microprocessor technology when they signed an agreement with Matsushita Electric Industrial Company to market the Fairlight CMI in Japan.[23] Ryrie explained in May 1987 how in the earlier stages of the company:

> [I]t was really just an ad hoc growth. Neither Peter nor myself had business management experience and so we were basically just running the company on whatever finances we could find, and when we sold an instrument the profits were ploughed right back into the company to help develop the next step.
>
> (quoted in Gilby 1987a, p. 52)

Some of the financial problems experienced by Fairlight Instruments related not to the manufacturing or the marketing of the Fairlight CMI but the distribution of the instrument and, specifically, the difficulties of selling to users in a geographically vast country.

After Bruce Jackson's initial attempts to demonstrate the Fairlight CMI to potential owners/users by flying prototypes of the instrument around the US, Fairlight Instruments set up branches in Los Angeles, New York, and Nashville. Vogel told me this was 'a financial disaster' (Vogel 2011a). Ryrie explained how:

> Our biggest market was Europe – specifically the UK and Germany. Japan was next. We had a lot of trouble in the US. In fact, the US has always been Fairlight's biggest trouble, which led largely to the downfall of the original company at the end of 1988. Our US subsidiary had lost almost $2 million in the previous two years, and the main company wasn't able to cover that during that post-crash period. We had three offices in the US, and the overheads there were extremely high. We've always found the US a very expensive place to sell into and support, because it's so physically large. It's very expensive to get out to all the population centres compared to, say, Europe.
>
> (quoted in Vail 2000b, p. 219)

Vogel drew attention to the company's lack of funding and explained that:

> We were reliant on sales to pay the wages and it was a horrendously expensive business. It was costing us something like $20,000 [AU] in components in each unit, so our market was rich pop stars. Our sales were good up to the last minute, but we just couldn't finance the expansion and the R&D.
>
> (quoted in Hamer 2005, p. 50)

The business strategy of manufacturing and selling a high-end instrument to a small number of wealthy pop and rock stars was becoming a problem. Vogel told me he would have liked to have developed a less exclusive product aimed at a larger market of users but 'lacked the capital and market penetration of their competitors' (Vogel 2011a). Like New England Digital, the profitability of Fairlight Instruments was impacted by the cheaper availability of new digital synthesizers, sampling keyboards, and rack-based samplers that were being developed in the mid-1980s. Their competitors were now US companies like E-mu and Ensoniq and Japanese companies like Akai, Casio, Korg, Roland and Yamaha.[24]

Launched in 1985 at a cost of £948, Akai's Midi Digital Sampler S612 has been described as 'the sampler that pioneered the low-cost market' (Gilby 1987b, p. 57).[25] In 1986, Casio launched the SK range of sampling keyboards with the SK-1, an 8-bit device that was reported to be the first digital sampler to sell more than one million devices across the world (Gilby 1987b). This figure contrasts with the small numbers of instruments being sold by Fairlight Instruments and New England Digital. In 1987, Casio launched the FZ-1. This was a 16-bit sampling keyboard costing £1,899 with a sample time of 29.1 seconds at 18 kHz or 14.5 seconds at 36 kHz, which could be expanded with additional RAM. The extended sample time led to one reviewer declaring: '[Y]ou could practically use the FZ-1 as a digital recorder for jingles, let alone as a sampler' (Jenkins 1987, p. 65). The Fairlight CMI Series III and the Synclavier II may have had technical advantages over many of the newer sampling devices. However, many potential users now judged these high-quality instruments to be too expensive compared to low-cost devices offering users similar amounts of sample time and sound quality levels.

The availability of 16-bit sampling on the Fairlight CMI Series III and the Synclavier II increased the perception that the use of 8-bit sampling on the CMI Series I and II could not reproduce sounds with satisfactory levels of realism. Yet, like J.J. Jeczalik, the lower-fidelity levels of the older technology were part of its appeal for Richard Burgess. He explained to me how 'the limitations of machines are a positive factor. I used to feel that about the Fairlight a lot. I used to think that the grungy, crunchy nature of the Fairlight was actually a cool factor. Even at the time I thought that' (Burgess 2011). The revival of interest in analogue synthesizers recently has been based on a nostalgia for their technological 'affordances'. Pinch and Trocco explain how '[f]or some people, the digital sound is too perfect, too clean, too cold – they long instead for the imperfections of the warm, fuzzy, dirty analogue sound' (2002, p. 319). Nick Prior observes that:

> [L]ike rock, electronic music has its own ideology of authenticity. This is, at first sight, less the romantic purity of unfettered human creativity and more an electronic hierarchy reconceptualised around given binaries – material over non-material, warmth over coldness, analogue imperfection over digital perfection.
>
> (2007)

The Fairlight CMI provided what users like Jeczalik and Burgess recognise as *digital imperfection*. When the Fairlight CMI-30A (Figure 2.3) was released in 2011 to celebrate the thirtieth anniversary of the Series I, the promotional literature described it as: 'a unique instrument, combining the latest technology with the look and feel of the original Fairlight CMI.

It achieves the classic Fairlight sound that defined the music of the eighties' (Fairlight 2011a). Unlike the digitally modelled PC version that is available, this version promises to bring 'a little eighties magic to the cold, hard digital world of 2011' (*ibid.*). Those who could afford a CMI-30A may have been nostalgic for the old digital 'warmth', imperfections, and limitations of hardware designed in the 1970s over the 'cold' digital 'perfection' of software packages being used for sampling in the twenty-first century.

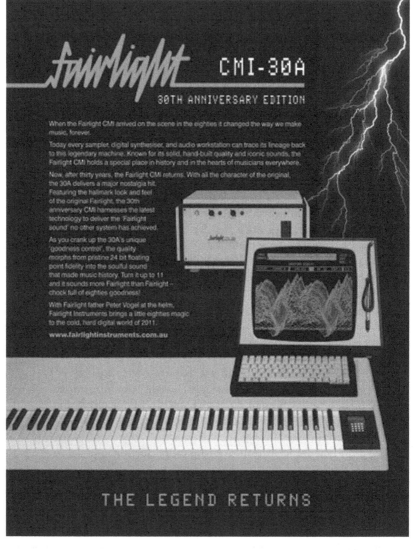

Figure 2.3 Fairlight CMI-30A 30th Anniversary Edition: The Legend Returns

Notes

1 The IIx was an 8-bit device with one second of sample time and a sample rate of 30.2 kHz. The Series III was a 16-bit device with sample rates of 44 kHz and 50 kHz (or 100 kHz in mono) and up to two minutes of sample time at 50 kHz (Fairlight 1986).

2 A more successful example of using digital synthesizers to reduce the number of keyboards for live performance was given by Steve Leonard, keyboard player in Los Angeles band, Cretones: 'I used to play with many more instruments, but I've replaced my B-3, Clavinet, Wurlitzer piano, and combo organ with a single Alpha Syntauri. If I were playing a Rhodes piano or a string machine, I would replace them with the Alpha, too' (quoted in Moog 1981, p. 77).

3 Ignoring the instruction manual, Jeczalik chose 'learning by doing' (Arrow 1962) or 'learning by using' (Rosenberg 1982). Rosenberg writes: '[I]n an economy with complex new technologies, there are essential aspects of learning that are a function not of the experience involved in producing the product but of its *utilisation* [original emphasis] by the final user' (p. 122).

4 For more on the life of session musicians/programmers, see Webley 1998a and 1998b: 'I first became an unofficial programmer in the early 1980s, when the first generation of LinnDrums hit town. . .It was a revolution, followed shortly by the arrival of the Fairlight, which needed a team of boffins just to switch it on and was better at drawing sine waves than it was at creating music. You also needed to be filthy stinking rich to own or hire one' (1998b, p. 28).

5 Jeczalik explained: 'I didn't really talk to him [Martin] about the Fairlight CMI. I had a bit of a bad session there. I had to rush back to the rehearsal studios because there was a leak. I had to leave the McCartney sessions and you don't do that. Someone else got the gig and I thought, 'Right. I know what I need to do now. I need to either work for Geoff or work on the Fairlight' so I decided to go on my own' (Jeczalik 2011.).

6 In the sleeve notes to *The Lexicon of Love*, Martin Fry wrote: 'A.B.C. were hell bent on making a record that would fuse two very different worlds. We loved Chic. We loved the Clash. We were through with matt and into gloss'. See Reynolds 2005 for more on post-punk/New Pop.

7 The inclusion of a sound to signify consumption is similar to the way Pink Floyd used the sound of a cash register on 'Money' (1973) from *Dark Side of the Moon* almost a decade earlier.

8 Discussing the stutter effect on Chaka Khan's 'I Feel For You', Arif Mardin stated: 'In the old days there was a sampler called the AMS and my finger slipped on the key so it became [sings] 'Chaka-chaka-chaka-khan' so we said, 'Let's keep it'. It was an accident (*laughs*)' (quoted in Burgess 2005, p. 282).

9 See Robertson 1983 and Bromberg 1989 for more on the legal action over songs on *Duck Rock* credited to McLaren/Horn. McLaren paid performers for session work but claimed to own the copyright and received royalties for songs he considered traditional and in the public domain. For discussions on issues relating to race, ethnicity, technology, and the appropriation of ethnic sounds as a form of exoticism, see Toop 1999, Born and Hesmondhalgh 2000, and Taylor 2007.

10 The instructions outlined the process of how DJs and MCs used turntables and a microphone: 'Two manual decks and a rhythm box are all you need. Get a bunch of good rhythm records, choose your favourite parts and groove along with the rhythm machine. Use your hands, scratch the record by repeating the grooves you dig so much. Fade one record into another and keep that rhythm box going. Now start talking and singing over the record with the

microphone. Now you're making your music out of other people's records' (quoted in Taylor 1988, p. 14).

11 Advertisements for the LM-1 focused on the authenticity of its sounds: 'Here's the most amazing rhythm machine *ever* [original emphasis] – the new Linn LM-1 Drum Computer from Linn Electronics. Amazing because it has real drum sounds – not synthesized noises, but *real* drums [original emphasis], digitally recorded and stored in memory' (Linn n.d.). An advert for the Oberheim DSX Digital Polyphonic Sequencer and DMX Programmable Digital Drum Machine in *Keyboard* magazine listed a number of design features including 'real drum sounds stored in digital memory' (Oberheim 1981).

12 The Fairlight CMI Series I had a loop function, but this was for repeating individual sounds by holding down a key on one of the piano keyboards rather than looping whole bars of music.

13 Jeczalik's willingness to admit he cannot remember clearly what happened during a recording session that took place in the early 1980s is helpful to the researcher as is his reluctance to claim the ability to construct particular loops was the result of anything other than 'goofing around'.

14 The DMX 15-80 Programmable Digital Delay Line/Harmoniser, DMX 15-80S Stereo Digital Delay Line/Harmoniser, and DMX 15-80SB Stereo Broadcast Delay Line were a series of microprocessor-controlled digital delay devices produced by Advanced Music Systems (AMS). By 1981, these could be fitted with a Loop Editing System.

15 When asked how MIDI came about, Dave Smith, who was president of Sequential Circuits in the early 1980s, said: 'Microprocessors were becoming standard in musical instruments. We figured out that it ought to be easy for them to talk to each other digitally. Everybody had their own digital interface, but none of them could communicate with each other' (quoted in Hamer 2005, p. 51). For more on the history and development of MIDI, see Moog 1983 and 1989.

16 The LinnDrum (1982) was Linn's follow-up to the LM-1 Drum Computer and also contained digital samples of acoustic drums. These were recorded with a sample rate of 35 kHz compared to the 28 kHz rate of the LM-1.

17 Langan describes Horn as: 'The first person I knew who had a great command of machines and he had this obsession about everything needing to be strictly in time. So, he was hell bent on using all the new machinery that was coming out which enabled him to achieve that' (quoted in Cunningham 1998, p. 272).

18 Timothy Warner drew attention to the combination of high-definition sounds and the 'poor' quality of the sampled sounds in his analysis of Art of Noise recordings. He writes: 'Sounds produced by the Fairlight Series II are often described as 'grainy': a quality which at the time was regarded as a deficiency but which nevertheless has a particular charm and character. . .This is especially noticeable when these samples are mixed with sounds of a higher resolution and sound quality. The Art of Noise often contrast the grainy samples of the Fairlight Series II with bright, clear sounds produced by synthesizers' (2003, p. 98).

19 After the release of *Who's Afraid of the Art of Noise?* (1984), Art of Noise planned to use 'Raiding the Twentieth Century' as the title of their next album. However, the album never appeared. Horn and Morley left, and Dudley, Jeczalik, and Langan continued the group as a trio.

20 The small print at the bottom of an advert published in *Studio Sound* magazine explained that this did not include the price of the music keyboard (Fairlight 1987). Other models with more RAM and additional disk storage were made available that year costing £49,950 and £112,000.

21 Ryrie stated: 'Until we looked at people using the Fairlight CMI in this application, I didn't realise just how little of the soundtrack is recorded during the original filming. All those footsteps, gunshots, and screams are all synced to film afterwards. This obviously takes weeks on a long film. Here, Jan Hammer is showing how he can turn an episode of *Miami Vice* around in *five days* [original emphasis]' (quoted in Gilby 1987a, p. 52).

22 Its managing director, David Hannay, explained: '[T]here was little point in Fairlight carrying on in the sampling market. Fairlight's business is not to compete with the high volume, mass-produced consumer products, and it never has been. It's to be leading edge with the latest technology, improve processing speed, and provide a high level of sophistication for the professional user' (quoted in Tingen 1996a, p. 53).

23 Matsushita manufactured products under the brand names of Technics, Panasonic, JVC, and Victor and became the Panasonic Corporation in 2008. The deal between Fairlight Instruments and Matsushita was reported in *Electronics Australia* magazine with the headline: 'Australian Synthesizer Cracks the World Market' (Williams, N. 1982).

24 For more on the development of Japanese synthesizers and keyboard instruments and the increase in sales and exports of synthesizers to the US in the early 1980s, see Doerschuk 1985.

25 An S612 cost £749 plus £199 for an Akai Sampler Disk Drive MD280 to store samples (Gilby 1985). It was a 12-bit device offering one second of sample time at 32 kHz or eight seconds at 4 kHz. As can be seen in adverts for the S612, Akai registered 'sampler' as a trademark (Akai 1985).

THREE

Technologies of Hip-Hop

The E-mu Emulator, SP-12, and SP-1200

In the 1970s Dave Rossum and Scott Wedge set up E-mu Systems to make analogue synthesizers. Pronounced 'Ee-myoo' and short for Electronic Music Systems, the company was initially named Eµ Systems.[1] Faced with financial problems at the beginning of the 1980s they decided to develop the Emulator, a keyboard instrument using the sampling technology that was of secondary importance to the Fairlight CMI designers. It is often referred to as 'the first affordable digital sampler'. Yet the Emulator and other sampling technologies designed by E-mu, like the SP-12 and the SP-1200 drum machines, were unaffordable for many potential users. Hip-hop is the genre of popular music most closely associated with digital sampling in the 1980s. However, hip-hop producers continued to use analogue technologies like turntables and magnetic tape to reproduce and repeat pre-recorded sounds. Some of the questions driving this chapter are: What technologies and sampling instruments were being used in the production of hip-hop in the 1980s? Who was using them and how were they being used? In this chapter, I explain how hip-hop became synonymous with the use and *misuse* of sampling technologies and continue to focus on the contingencies that occur during the design of music technologies. I show how instruments are not only used in ways unimagined by their designers but in ways that are perceived to conflict with their principles, values, and marketing strategies.

Following the Designers and the Instrument: E-mu Systems and the Emulator

Like Peter Vogel and Kim Ryrie at Fairlight Instruments, Dave Rossum and Scott Wedge were high school friends who began designing synthesizer technologies in a domestic environment. Rossum started E-mu Systems in 1970 and rented a house at 625 Water Street in Santa Cruz, California the following year.[2] He developed an interest in synthesizers as an undergraduate student at the California Institute of Technology (Caltech). As a graduate student of Microbiology at the University of California in Santa Cruz (UCSC), he was introduced to a newly acquired Moog modular system, The Synthesizer 12, in the University's Electronic Music Studios. With friends from Caltech, Steve Gabriel and Jim Ketcham, Rossum built a small prototype of a synthesizer called the Black Mariah with a tin foil keyboard; a second prototype called the Royal Hearn was built in the summer of 1971 with fellow UCSC students, Paula Butler, Marc Danziger, and Mark Nilsen. Wedge, who studied at the University of California in Berkeley before dropping out, joined the company after suffering a back injury in a skydiving accident. They developed a synthesizer with a three-octave keyboard called the Eμ 25, which was modelled on The Synthesizer 12 and ARP 2600 synths. In November 1972, Rossum and Wedge formed E-mu Systems as a legal entity and moved to the City of Santa Clara in Silicon Valley. They began manufacturing and selling their own Modular system with a five-octave monophonic keyboard, which was launched in 1973. There was no fixed price, but the first one cost $4,000.[3] Over the next few years, they experimented with using microprocessors to control synthesizers and, in 1977, were commissioned by Peter Baumann of Tangerine Dream to build the Audity Level I System. This was a smaller, more portable 'workstation' that was launched in May 1980 at the Audio Engineering Society (AES) convention in Los Angeles. The projected cost was $50,000 but ended up closer to $70,000. Realising this was too expensive and faced with a dispute with Dave Smith of Sequential Circuits over royalties owed for work on the Prophet-5 synthesizer, Rossum and Wedge moved into the design of digital sampling instruments and focused on developing a cheaper technology than the Fairlight CMI by using less memory.[4]

At the AES show in May 1980, Rossum and Wedge saw demonstrations of the Fairlight CMI as well as the Publison DHM 89 B2 and LM-1 Drum Computer, which they had worked on with Roger Linn. Rossum explained why they were unimpressed by the synthesizer/sampling instrument designed in Australia:

> Wedge, [general manager] Marco Alpert, and [head technician] Ed Rudnick had been talking on the drive back from the show and thought that the Fairlight [CMI] had one and only one good feature – sampling. We had also seen a Publison Digital Delay that had a capture mode, and the captured (sampled) sound could be played with a control voltage/gate-type synthesizer keyboard. The guys came to me with their ideas, and we had the need for a new MI [Musical Instrument] product quickly to replace the lost Sequential revenue stream.
>
> (quoted in Abildgaard 2012)

The Fairlight CMI used a separate Central Processing Unit (CPU) and Random Access Memory (RAM) for each of its eight voices or samples. Rossum realised there was a less expensive way of doing this and wanted to use one CPU to deliver an eight-voice polyphonic instrument. In order to increase the available memory, Rossum found a solution using Direct Memory Access (DMA) chips and FIFO data buffers.[5] He described the overall design process as 'revolutionising the state of the art – building what was in my mind, not duplicating something that I'd seen' (*ibid.*), yet also admitted to wanting to emulate competing technologies like the Fairlight CMI that used digital sampling:

> We knew that all of these products were fairly hot, and of interest to most musicians, most of whom couldn't afford them. Being the sort of people who didn't mind borrowing other people's ideas, we said, 'It sounds like this digital sampling idea is ripe. Someone should come in and do it right'.
> (quoted in Vail 2000c, p. 221–222)

The move by E-mu Systems, from designing synthesizers like the Audity towards developing the first dedicated sampling keyboard and a more affordable instrument than the Fairlight CMI, was born out of financial necessity. Their decision to focus on digital sampling demonstrates again the contingency of the instrument design process.

Figure 3.1 E-mu Emulator

E-mu began work on the hardware and software for the Emulator in June 1980 and launched the prototype at NAMM's Winter Market in February 1981. With a four-octave keyboard and a 5¼ inch disc drive for storing sounds, it offered users two seconds of sample time. Ten diskettes were supplied. Eight contained pre-programmed sounds and two were blank for users to record their own sounds, or what would subsequently be referred to as sampling. The Emulator, though, might not have been called the Emulator. Wedge explained:

> Whenever we do a project, we have an in-house name for it. Then, as we get closer to the time that it goes to market, we go through a formal process of actually naming the product. The in-house product name for the Emulator was the 'Sampler'. For us, that was kind of a pun between Nyquist's sampling theorem – which is an obscure piece of mathematics that underlies the whole genre – and the Whitman Sampler, a box with a whole bunch of different flavours of chocolates in it, because this was an instrument that could have a whole bunch of different sounds.
>
> (quoted in Vail 2000c, p. 224)

In December 1980, musicians were invited to E-mu's base, which was now a 'commercially zoned house' (E-mu 2015) at 417 Broadway in Santa Cruz. They tested the prototype by sampling instruments and checking their fidelity levels. The Emulator was also tested by Rossum's then girlfriend who recited the same nursery rhyme Thomas Edison had recorded with a tinfoil phonograph more than a century earlier:

> [T]he first loop was [Rossum's future wife] Karen speaking into the instrument, saying 'Mary had a little lamb'. And I [Rossum] could simply hold down the key and it would play 'Mary had a little little little little lamb'. The next loop I made, after Karen left, was me peeing in the toilet adjacent to the lab. It made it sound like I had the world's largest bladder.
>
> (quoted in Abildgaard 2012)

A fascination with sampling the sounds of the lavatory over the laboratory extended to E-mu's marketing campaign for the Emulator, which imagined users would record and loop sounds of their own. As well as promoting the keyboard as a way of imitating the sounds of acoustic instruments, E-mu encouraged users to sample 'the sounds of everyday life'.

Imagine . . .

Imagine being able to play any sound you can hear—polyphonically.
Imagine a computer-based instrument that can record any sound into its digital memory—either live from a microphone or from a line level source—and then allows you to play that sound at any pitch over the range of its keyboard. With up to eight note polyphonic capability. An instrument whose split keyboard allows simultaneous control of two independent sounds. Any sounds. Instruments. Voices. Sound effects. Animals. Machines. Anything.

Imagine having realtime control over any sound.
Imagine being able to sustain a sound indefinitely—regardless of its original length. Or using performance oriented effects wheels to add frequency modulation to any previously stored sound. Turn a single trumpet note into a brass section. Add true vibrato to a grand piano. Play barking dogs. Polyphonically. With pitch bend.

Imagine an instrument that's incredibly simple to use.
Imagine an instrument that requires no special programming skill. An instrument that allows you to store and

recall sounds in a matter of seconds with its built-in floppy disk drive. An instrument that is completely self-contained and portable, with no bulky external computer or CRT terminal to carry around.

Imagine an advanced digital instrument that you can afford.
Imagine this instrument available in two, four, and eight voice versions with suggested prices ranging from $6000 to under $10,000.

On February 6 at NAMM E-mu Systems will introduce the Emulator polyphonic digital keyboard instrument.
Then you won't have to imagine.

E-mu Systems, Inc.
417 Broadway, Santa Cruz, CA 95060
(408) 429-9147

Come to booth 1435 at NAMM. Hear the future.

Figure 3.2 'Imagine. . .' advertisement (*Contemporary Keyboard*, February 1981)

Like initial adverts for the Fairlight CMI, which presented it as a revolutionary new technology that enabled users to create 'tomorrow's music today', E-mu wanted potential owners and users of the Emulator to 'hear the future' (Figure 3.2). The approach to marketing focused on futurism, fidelity, and fun. An advert published in the May 1981 issue of *Contemporary Keyboard* magazine (Figure 3.3) contained a pun on the US word for a failed product:

> Play a Turkey. Or a dog. Or violins, drums, voices, sound
> effects, machines, or, in fact, anything. Not synthesized
> simulations but the actual sounds. With the E-mu Systems
> Emulator, *any* sound [original emphasis] you can hear can
> be digitally recorded and then played back at any pitch
> over the range of its keyboard. . .
>
> (E-mu 1981)

In the UK, an advertising slogan produced by E-mu's distributor, Syco
Systems, read 'From farts to filharmonics [*sic*]' (Syco Systems 1981).
E-mu were keen to stress that any sound could be used in the process
of making music and were more playful in their approach to marketing
than the focus in adverts for the Fairlight CMI on 'ANY type of music' or
'ANY type of sound'. As with the emphasis in adverts for the Synclavier
and Linn LM-1 Drum Computer on reproducing 'real sounds', E-mu's
campaign also employed a discourse about authenticity and unmediated
sounds. An advert in *Keyboard* magazine in October 1982 read:

> Finally, there's nothing standing between you and
> the sounds you want. *Any* sound you want [original
> emphasis]. Instruments. Voices. Sound effects. Animals.
> Machines. Anything. Sounds that sound real because
> they *are* real [original emphasis] (E-mu 1982).

Claims around realism and fidelity were central to the attempts by E-mu
to sell a sampling keyboard that could be used to reproduce sound as a
'vanishing mediator' (Sterne 2003, p. 283). According to its adverts, the
Emulator could be used to reproduce 'actual' sounds or 'real' sounds rather
than produce digitally synthesized sounds. Yet these were not 'actual' or
'real' sounds but digitally recorded samples of 'actual' or 'real' sounds.

Despite offering users the opportunity to sample 'the sounds of everyday life' and the lower cost of buying a sampling instrument – $9,995
[US] – sales of the Emulator were slow. Starting in July 1981, E-mu's
business plan was to sell five instruments a month. According to Rossum,
'We sold about 20 of our first units, but sales just hit the wall at the end
of 1981' (quoted in Vail 2000c, p. 225). The poor sales of the Emulator
were attributed to technical issues. As E-mu's Director of Marketing,
Alpert concluded that users were unsure how to use a musical instrument
that enabled any sound to form part of a recording:

> People didn't know what to make of it. People who had
> Fairlights knew, but there weren't that many people
> who had Fairlights. And there wasn't a paradigm yet
> that everyone was familiar with. It had a slow build. It

Figure 3.3 'Play a Turkey' advertisement (*Contemporary Keyboard*, May 1981)

took about a year, and it was really our introduction of a sound library that you could get along with it that helped out. So, you could take it out of the box, put a few discs in, and have a bunch of useful sounds right then and there, rather than having to go out and figure out how to do it yourself.

(quoted in Milner 2009, p. 320)

To make it more affordable and user-friendly, E-mu lowered the price of the Emulator to $7,995 [US] in January 1982 and increased the number of diskettes with pre-programmed sounds to twenty-five. However, the inclusion of a sample library of pre-set sounds was ignored by at least one user in the world of hip-hop who discovered the keyboard instrument could be used to sample the sounds of musicians and, more specifically, the sounds of drummers on old vinyl recordings. Understanding African-American musical practices, such as the focus on rhythm, repetition, and rupture (Rose 1994), is important for understanding the technological practices that shaped the production of hip-hop. I therefore want to re-introduce the concept of 'relevant social groups' from the field of Science and Technology Studies (STS) and the social construction of technology (SCOT) approach to examine how hip-hop producers used sampling technologies and how sampling became associated with reproducing sounds from pre-existing recordings.

Following the Users (and Non-Users) of the Emulator: The Marley Marl Moment

The first Emulator owner was Stevie Wonder, who visited the E-mu stand at NAMM in February 1981. In an example that illustrates the gap between promises contained in magazine adverts and the experience of using prototypes when they are launched, he began to sample sounds on the Emulator in ways that the designers felt highlighted its limitations. Wedge explained how the Motown artist and Fairlight CMI owner

> . . .walked up to the instrument, sort of hugged it to get the feel of it, and then started playing it. . .Stevie sampled his voice into the Emulator and played it back on the keyboard. That drove us all crazy because we knew that voice didn't work very well on it. Voices ended up sounding funny. 'Munchkinised' was what we called it. We thought there were much more interesting things to sample. To top it off, when Stevie sang into the microphone for the sample, it really overloaded the inputs and distorted the signal. It was a bad sample and a bad example, but when he played it, I guess it was just enough of a mindblower to turn him on to it.
>
> (quoted in Vail 2000c, pp. 224–225)

Daryl 'the Captain' Dragon (of the husband-and-wife duo Captain & Tennille) had been promised ownership of the first Emulator, but Stevie Wonder was thought to be a more high-profile customer and Dragon had to settle for serial number 002. His initial use of the Emulator included the creation of sound effects such as sleigh bells and reindeer hoofs at a Christmas concert in Los Angeles with the Glendale Symphony Orchestra (Vail 2000c). For those who could not afford to buy an Emulator, the other way to become a user was to hire one. Paul Hardcastle chose this option when he used an Emulator on '19' (1985), a number one single in the UK with a stuttering effect ('N-N-N-Nineteen').[6] The user of the Emulator I want to discuss was not an owner either: Marley Marl worked as an assistant to the producer Arthur Baker and this semi-professional relationship gave him access to sampling technologies like the Fairlight CMI and E-mu Emulator.[7]

When digital synthesizer/sampling technologies like the Fairlight CMI first became available in the late 1970s and early 1980s, record labels and producers made recordings of hip-hop performances using the skills of DJs and musicians to replay breakbeats (or breaks).[8] On 'Rappers Delight' by Sugarhill Gang (1979), musicians replicated the sounds of pre-existing recordings that the producer, Sylvia Robinson, wanted to use. A member of the house band at Sugarhill Records, drummer Keith LeBlanc, told me:

> The DJs would come in with a bit of a record and we had an arranger named Jiggs Chase who would write an arrangement of it and then we would play [it]. There [were] no samplers around and we would have a chart written and I'd add little things to it. Everyone would add little things to it.
>
> (LeBlanc 2008)

LeBlanc described how the technical process of making records for the label changed with the introduction of synthesizers and drum machines containing samples. Rather than these instruments replacing the house bands, the same musicians programmed and played them. The use of a synthesized keyboard sound can be heard on Grandmaster Flash and the Furious Five's 'The Message' (1982). LeBlanc supplied some technical information about its production when I asked him whether he played on the recording of 'The Message' that became a hit single: 'I played on a [different] version of 'The Message'. There were two versions cut. I didn't play on 'White Lines' either. Reggie Griffin did the drum machine on that. A[n Oberheim] DMX' (*ibid.*). As the use of synthesizers and drum machines with samples changed the sounds of hip-hop in the early 1980s, hip-hop producers also began to sample the sounds of acoustic drums from vinyl recordings.

The discovery that digital samplers could be used to record and re-use sounds from pre-existing recordings is associated with hip-hop producer, Marley Marl. In an interview with *The Source* magazine in 1991 he explained this was a recording studio accident:

> One day in '81 or '82 we was doin' this remix of a
> Captain Rock record for [indie label] Nia. I wanted to
> sample a voice from off of this song with an [E-mu]
> Emulator and, accidentally, a snare went through. At
> first, I was like, 'That's the wrong thing,' but the snare
> was soundin' *good* [original emphasis]. I kept running
> the track back and hitting the Emulator. Then I looked
> at the engineer and said, 'You know what this means?!
> I could take any drum sound from any old record, put
> it in here and get the old drummer sound on some shit.
> No more of that dull [Oberheim] DMX shit'. That day
> I went out and bought a sampler, a little cheap bullshit
> sampler I still use to this day. 'Marley's Scratch' was the
> first record to use sampled drums, but [the innovation]
> really got noticed [when it appeared] on 'The Bridge' and
> 'Eric B is President'. I had made my own patterns with the
> [The Honey Drippers'] 'Impeach the President' snare and
> kick. That was the shit, I was excited.
>
> (quoted in Nelson 1991, p. 38)[9]

What might be referred to as 'the Marley Marl moment' is problematic because it suggests the use of digital samplers in hip-hop was becoming widespread in the early 1980s when, in fact, they were not widely adopted until the arrival of less expensive devices like Akai's S900, Casio's SK-1, and E-mu's SP-1200 in the mid-to-late 1980s. Joseph Schloss suggests Marley Marl's discovery 'almost immediately ended the era of live instrumentation' (2014, p. 35), but the production of hip-hop in the early 1980s involved a diverse range of musical/technological practices. The recording of performances using acoustic and electric instruments such as guitars, bass, and drum kits did not come to a sudden halt: the types of musical instruments changed as musicians began learning how to use (digital) synthesizer and sampling technologies alongside more familiar instruments. Hip-hop producers realised they could sample drum sounds and breaks from pre-existing recordings but were limited by the two-second sample time of the Emulator. For Tricia Rose, the consequence of this accidental discovery was that 'real drum sounds could be used in place of simulated drum sounds' (1994, p. 79). However, what Marley Marl described was a preference for a particular sound: the digital samples of acoustic drums on pre-existing analogue recordings rather than the digital samples of acoustic drums stored in the memories of drum machines like the Oberheim DMX.

While instrument designers like Linn and Oberheim advertised the importance of 'real' sounds that had been digitally sampled during the making of their drum machines, the ownership and use of these instruments was restricted by price: Linn's LM-1 Drum Computer cost $5,000 [US] when it was released in 1979 and the Oberheim DMX cost $3,000 [US] when it became available in 1981.[10] As an intern working in a Manhattan recording studio, Marley Marl had access to an Emulator and an Oberheim DMX but was critical about the timbre of the digitally sampled sounds stored in the memory of the DMX. Other hip-hop producers became key users with one adopting the name of the instrument. David Reeves, a bassist, guitarist, and DJ for Kurtis Blow, became known as Davy DMX. When asked in an interview why he named himself after the drum machine, he replied:

> It was the hottest thing out back then. There was the Linn [LM-1 Drum Computer and LinnDrum] but once the DMX hit, you just had to mess with it. It was one of the first machines that came out that had a decent drum sound.
>
> (quoted in Coleman 2013, p. 148)

In the early 1980s, digital samples used in hip-hop were more likely to be pre-programmed sounds contained in the Oberheim DMX or the sample library of the Fairlight CMI (as on Afrika Bambaataa's 'Planet Rock', for example) than sounds recorded externally using digital synthesizer/sampling instruments or sampling keyboards like the Emulator. While Davy DMX enjoyed the sampled sound of acoustic drums stored in drum machines and Marley Marl the sampled drum sounds on vinyl records, other producers favoured drum machines with sounds produced using analogue circuitry like the Roland TR-808.[11]

Until samplers became more affordable in the mid-to-late 1980s, reproducing the sounds of pre-existing recordings occurred mainly through the use of analogue technologies: DJs used vinyl on turntables to repeat breaks and producers cut and spliced magnetic tape in recording studios. Steve Stein, an advertising executive and aspiring DJ, and Douglas Di Franco, a sound engineer, entered a Tommy Boy Records remix contest in 1983. They won and the results, which owed as much to the break-in records of Buchanan and Goodman as the hip-hop turntablism of Grandmaster Flash, were released as 'The Payoff Mix' (1984) under the name of Double Dee & Steinski. The mix was made using turntables and tape machines. Stein recalls: 'Douglas would listen to what we had and then figure out what we needed, so he'd record a couple of pieces from the record onto a two-track tape, cut them together, then vari-speed that tape so it was in sync with the larger eight-track tape' (quoted in Hsu 2008). LeBlanc explained how he used drum machines and magnetic tape to create 'No Sell Out' (1983), which included excerpts from speeches by Malcolm X:

> I programmed it with an Oberheim DMX and E-mu
> Drumulator. I edited out bits of Malcolm X's voice and
> put them on 2-inch tape and flew them in. It was before
> samplers. It was a lot harder to do. The tape machine
> was hit and miss but you could play it like an instrument
> after a while. Samplers made it easier.
>
> (LeBlanc 2008)

In a compilation released by Sanctuary Records in 2005, the liner notes
describe how this 'celebrated electro classic from 1983 was one of the
first to use new sampling technology, repeating sections of Malcolm X
speeches into a heavy beatbox groove' (Sanctuary 2005). However, any
digital sampling technology used in the making of 'No Sell Out' related
to the construction of the rhythm track rather than the repetition of
Malcolm X's words. Until the arrival of samplers and sampling drum
machines with enough memory to store samples of complete breaks from
vinyl recordings, the use of analogue technologies like magnetic tape and
turntables remained the primary way of creating the appropriation-based
approach to record production that was central to hip-hop's aesthetic.

Following the Instruments: The Drumulator, the Emulator II, and SP-12

E-mu's decision in 1982 to re-introduce the Emulator by reducing the price,
adding new features such as multi-sampling, and increasing the number of
sounds in the sample library was successful. They sold seventy-five keyboards at NAMM that year and sales of the instrument remained steady.
Twenty-five instruments were produced each month until the product was
discontinued before the introduction of the Emulator II in 1984. Just as the
Emulator was designed as a cheaper alternative to a more expensive syn-
thesizer/sampling instrument, E-mu produced a less costly drum machine
containing digital samples to compete with Linn's LM-1 and LinnDrum
products. The suggested retail price of the LinnDrum was $2,995 [US]
(Linn 1982). E-mu introduced the Drumulator at NAMM in January 1983
with a retail price of $995 [US] and the marketing emphasised its 'afforda-
bility'. An advert in the February 1983 issue of *Keyboard* magazine stated:
'[Y]ou have a digital drum computer that would be amazing value at
$1990.00. But what's even more amazing is that for $1990.00 you would
get something that you probably wouldn't expect. Two Drumulators.'
(E-mu 1983a).[12] The owner's manual explained to users:

> The Drumulator is a rhythm/drum machine that fea-
> tures twelve digitally recorded drum sounds stored on
> computer chips, and extensive solid-state recording

capabilities. You may record up to thirty-six individual rhythm patterns (called *segments*), and then combine these segments in just about any order imaginable to create up to eight songs.

(E-mu 1983b, p. 4)[13]

Like the Linn LM-1, the Drumulator did not have a user sampling function and some users were frustrated with the limited number of pre-programmed sounds. Two of these users wanted to increase the number of available drum samples. Peter Gotcher and Evan Brooks, who had been students of electrical engineering and computer science at University of California Berkeley, contacted E-mu for advice about how to re-design the instrument. To their surprise, Alpert, Wedge, and Rossum were encouraging and sold them an Emulator to sample their own drum sounds.[14] The duo began designing chips with drum sounds to sell to Drumulator users – these included electronic drum sounds, Latin and African percussion, and a heavy metal rock drum set – and started a company called Digidrums. These users of a drum machine containing samples of acoustic drums became the designers of sample libraries of drum sounds and, later, Digital Audio Workstations (DAWs).[15]

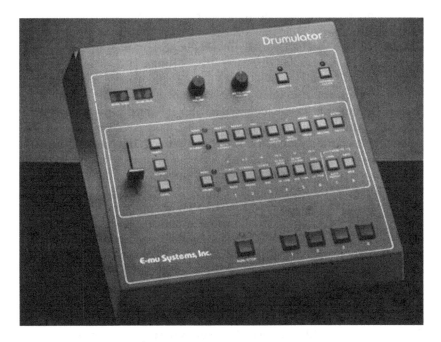

Figure 3.4 E-mu Drumulator

Figure 3.5 E-mu Emulator II

For E-mu, the Emulator was important to the development of the company because it was their first instrument to be distributed and sold in music retail stores. The Drumulator was important because of its commercial success with sales of nearly 10,000 units in a two-year period (Keeble 2002).[16] Wedge explained: 'What actually made E-mu wasn't the Emulator. It was the Drumulator. But the Emulator made the Drumulator, and the Drumulator then made the Emulator II, which really was our first truly successful instrument' (quoted in Vail 2000c, p. 225). This was not, though, a straightforward story of business growth and financial success. The Emulator II was introduced at NAMM in January 1984, but Rossum admitted: '[W]e showed a prototype that was barely completed and was nowhere near ready for production' (quoted in Grandl 2015b). With no revenues from the Emulator, which had been discontinued, E-mu licensed a product called the ddrum from a company called Clavia in Sweden and launched the E-drum Digital Percussion module – an electronic drum pad using digitally recorded drum sounds. Rossum described it as 'a complete catastrophe. It was plagued by reliability problems' (*ibid.*). With poor sales, E-mu were forced to make redundancies among its workforce and a deal agreed with their distributor in the UK, Syco Systems, enabling the company to stay in business.[17] When E-mu finally started to distribute the Emulator II, they sold more than 3,000 units over a three-year period.[18] It was an 8-bit device with a sample rate of 27.777 kHz and 17.6 seconds of sample time (Keeble 2002).[19] Its fans celebrated the realism of its sounds – Paul Wiffen wrote about how 'the EII [Emulator II] was the first sampler I ever came across which could even get close to a piano' (2000, p. 264). But with a price tag of $7,995 [US], there is little evidence that the Emulator II or E-mu's next commercially available product – a drum machine – was widely used in the production of hip-hop or among social groups other than rich rock and pop stars.

E-mu Systems SP-12 Twelve-Bit Sampling Percussion System was not always called the SP-12. In February 1985 at Frankfurt Musikmesse trade fair, E-mu launched the Drumulator II, which combined features from the Drumulator and the Emulator II. When it was distributed later that year, they had changed its name to the SP-12. In the UK, the recommended retail price was £2,995 plus an additional £500 for a Turbo version (Wiffen and Scott 1985). In the US, it was $2,745 for the standard version and $3,550 for a Turbo version (Oppenheimer 1986). A 12-bit device that enabled users to sample their own drum sounds or other sounds that could be used for percussion, the standard version included 1.2 seconds of sample time. The Turbo version included five seconds, though this was distributed across two memory banks with a maximum sample time of 2.5 seconds each. With a similar interface to the Drumulator, added to the eight velocity-sensitive programming buttons were sliders for adjusting the volume or pitch of any of the sounds. Now describing the company as 'pioneers in affordable professional

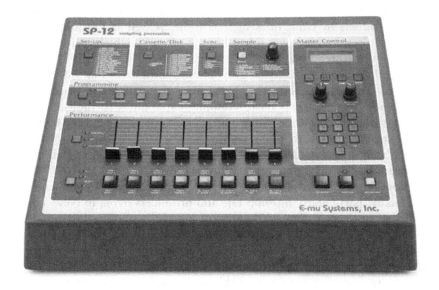

Figure 3.6 E-mu SP-12 Twelve-Bit Sampling Percussion System

sampling technology', E-mu argued that the SP-12 set a 'new standard of fidelity for digitally sampled drum machines' (E-mu 1985). Reviewers agreed. Paul Wiffen and Annabel Scott wrote that 'next to a Linn 9000, the SP-12 makes its competitor sound dull, muffled, and uninspiring. *And* [original emphasis] the factory sounds are refreshingly modern: deep, powerful toms, a sharp, clicky bass drum, a good selection of electronic kit sounds, plenty of realistic ethnic percussion' (Wiffen and Scott 1985, p. 45).[20] To store sampled sounds, the SP-12 was designed to be used with a JL Cooper MIDI Disk drive and could also be used with a Commodore 1541 disc drive (Oppenheimer 1986). Applying the discourse used to sell synthesizers to the marketing of digital drum computers, E-mu stated: 'Virtually anything you can imagine can be sampled into battery backed up memory' (E-mu 1986). Joseph Schloss reports how 'hip-hop artists were soon using the machine to sample not their own drumming, but the sound of their favourite recorded drummers, such as Clyde Stubblefield from James Brown's band or Zigaboo Modeliste of the Meters' (2014, p. 35). However, until the launch of a later version, the SP-12 does not appear to have been widely used in the making of hip-hop records.

Following the Users/Non-Users of the SP-12: Boogie Down Productions (BDP) and Beastie Boys

To understand if, and how, the E-mu SP-12 and sampling drum machines were being used by hip-hop producers in the mid-1980s, we need to know who had access to these technologies and whether they had, in reality, become more affordable. The introduction of new sampling products by companies such as E-mu as well as Ensoniq, Akai, and Casio resulted in lower prices but the extent to which their use had become widespread may have been exaggerated. In 1985, Bob Moog wrote: 'Like the democratised polyphonic synthesizers that influenced the musical instrument market because of their attractive prices, the [Ensoniq] Mirage brings basic sampling capabilities to thousands of eager musicians at a price where no such instrument had existed before. And costs continue to plummet' (p. 46).[21] Affordability is relative, though, and applies differently to 'relevant social groups' including African-Americans living in less affluent inner-city areas in the 1980s. Schloss has explained how these technologies were still too expensive for many, particularly hip-hop DJs and producers living in New York housing projects.[22] The availability of cheaper sampling instruments like E-mu's SP-12 and Ensoniq's Mirage made the digital sampling of external sounds more accessible to a larger number of users. However, the argument that this represented a form of democratisation because the technologies were now available to everyone is too simplistic and not supported by empirical evidence.[23]

Interview material from hip-hop musicians such as KRS-One (Kris Parker) from Boogie Down Productions (BDP) suggests ownership of the SP-12 was still uncommon within specific socio-cultural communities between 1986–1987. He highlights this when describing how 'South Bronx' from *Criminal Minded* (1987) was produced:

> I performed the verses for Scott, he played 'Funky Drummer' and started in on the song, and it blew his mind. So, we ran over to Ced-Gee's house and were like: 'Yo, Ced, we need that SP-12'. Keep in mind that at that time Ced-Gee was the only person in the Bronx with an SP-12, and he was the absolute man. So, he lent us the sounds, the kick, the drum, the snare, the hi-hat. Scott took his records over to Ced and Ced sampled them and made the beat for 'South Bronx'. Scott did the drums and Ced chopped it up.
>
> (quoted in Coleman 2007, p. 82)[24]

It is impossible to know *how many* SP-12s existed in the Bronx at this time, but the above quote suggests they were scarce as hip-hop users borrowed them from owners who possessed significant cultural capital (Bourdieu 1984) and social capital (Bourdieu and Wacquant 1992). What was also in limited supply was specific knowledge about how to sample sounds from pre-existing recordings, which is why a producer like Ced-Gee was also relied upon to programme the instrument. With only 1.2 seconds available to users of the standard version of the SP-12, *Criminal Minded* contained samples of only a short length. On 'Poetry', the scratching of vinyl on turntables is accompanied by snatches and shrieks from a James Brown recording. 'Dope Beat' features guitar riffs from AC/DC's 'Back in Black'. The limited amount of sample time on the SP-12 meant users were reproducing individual drum sounds such as those from Clyde Stubblefield's solo on 'Funky Drummer', but complete breakbeats could not yet be sampled and looped.[25]

Released in 1986, the Beastie Boys' *Licensed to Ill* contained drum breaks from recordings such as Led Zeppelin's 'When the Levee Breaks'. Some journalists assumed that pre-existing sounds on the album were reproduced using an SP-12. Angus Batey, for example, writes:

> Musically, *Licensed to Ill* is basic. Like much of the hip-hop of the time, it relies on a selection of beats concocted on machines like the legendary SP-12, a drum machine that allows the programmer to construct original percussive patterns using sampled drum sounds. For instance, 'Rhymin & Stealin', the opening track, uses a mixture

of deck techniques and drum machine programming to turn a sequence of sampled John Bonham drums into a slow, loping, lazy hip-hop rhythm. Still a relatively new tool in the mid-1980s, the SP-12 was behind most of the major stylistic advances in hip-hop music prior to the advent of cheap samplers with long sample times (which, oddly, merely facilitated a return to the 'live' sounds a DJ could create by mixing records on a pair of turntables), and its distinctive sound underpins much of the rest of the album.

(1998, p. 40)

In fact, the drum sounds from 'When the Levee Breaks' were repeated by recording and looping them using magnetic tape. MCA of the Beastie Boys explains:

On *Licensed to Ill*, we didn't even have any samplers. So, the stuff that's looped, we actually made tape loops. We'd record 'When the Levee Breaks' beat onto a quarter-inch tape, and then we'd make the loop and that tape would be spinning around the room, dangling on mic stands, going around in a big loop. And then, in order to layer that with something else, we'd have to actually synch it up, physically.

(quoted in Brown 2009, p. 45)

The assumption is that the Beastie Boys used an SP-12 in the making of *Licensed to Ill* because it became available in 1985 and the album was released the following year. However, those involved in the engineering of the album were relying on established skills that had been part of the process of record production since the 1960s as well as new sets of skills needed to programme digital reverb and drum machine technologies that were emerging in the 1980s.[26] The use of magnetic tape to loop and repeat excerpts from pre-existing recordings was still an important practice in hip-hop in the mid-1980s.

While E-mu's SP-12 and other sampling instruments were used in the production of hip-hop at this time, DJs continued to be deployed to incorporate pre-recorded sounds into new recordings. On *Paid in Full* (1987) by Eric B. and Rakim, Eric B. scratched in brass sounds using turntables and vinyl, the engineer, Patrick Adams, produced rhythms using the pre-set sounds of an analogue drum machine, and Marley Marl sampled sounds from pre-existing recordings.[27] Rakim explains the process of making 'I Ain't No Joke', the first track on the album:

> That sample was just another James Brown record that
> I used to rhyme off. At first, we were going to sample
> more of it, but then we decided to leave Eric just scratch
> [the horn riff] in. . .With the drum programming on the
> album, our engineer Patrick Adams did a lot of that. I'd
> just basically take my break beats and ideas in, and he'd
> sample it up and put the [Roland TR-]808 on it.
>   (quoted in Coleman 2007, p. 206)[28]

Musicians using electric instruments were also employed when performances
on recordings could not be digitally reproduced. On the title track of the
album, Eric B. and Rakim used the bass line from 'Don't Look any Further'
by Dennis Edwards (1984). Rakim recalls: 'With that track, I always used to
rhyme off that. . .in the park. Eric put that beat up under the bass line. I think
that was Patrick Adams replaying the bass line' (quoted in Coleman 2007,
p. 208). Rather than a revolutionary process in which the introduction of
digital samplers and sampling drum machines like the SP-12 changed the
ways in which hip-hop recordings were made, records with pre-existing
sounds were produced using sampling technologies, analogue technologies like tape and turntables, and electric instruments. As devices with more
memory, more sample time, and lower prices became available, digital
sampling became the primary way of reproducing sounds from pre-existing
recordings, as part of a more gradual process of socio-technological change.

Figure 3.7 E-mu SP-1200 Sampling Percussion System

Discontinuing the Instrument: The SP-1200 and the Acquisition of E-mu

The designers and marketers at E-mu realised users of the SP-12 were disregarding its built-in drum sounds and wanted more RAM to sample their own sounds. Alpert states: 'Everyone was ignoring the built-in sounds and going, 'We want more memory to do it ourselves'' (quoted in Milner 2009, p. 331). E-mu released an updated version, the SP-1200, in 1987. Advertisements referred to 'the formidable and ever-expanding SP-1200 library of sounds' (E-mu 1987). The user's manual encouraged users to sample acoustic and electronic drum sounds as well as sounds from records, CDs, and tapes. Where the SP-12 required an external drive for storing sampled sounds, the SP-1200 included an internal drive and came with five 3½ inch floppy disks containing pre-programmed sounds. Available in the UK for £2,199 (Mellor 1987), E-mu's adverts promised 'a full 10 seconds of sampling time' with a rate of 26 kHz. However, as sample time was distributed across four banks that each stored eight sounds, the maximum length of samples was still only 2.5 seconds. Its use by hip-hop producers was the result of contingency rather than E-mu's marketing strategy. Alpert explains:

> I designed the user interface for the SP-1200, and while I would like people to think I was prescient as to think it would be a cool tool for rap and hip-hop people, it was totally by accident. None of us had any idea that what we were doing would be used in that particular way. But people loved that interface. The SP-1200 was very approachable and intuitive and immediate. And then we couldn't even kill it.
> (quoted in Milner 2009, p. 332)

Adopters of the SP-1200 in hip-hop such as Pete Rock and Hank Shocklee valued its 12-bit fidelity levels at a time when 16-bit instruments like the Fairlight CMI Series III and the Casio FZ-1 were available. They also discovered ways of overcoming the 'affordances' or technological constraints of the instrument. Fixes were developed so that, where possible, users could sample an excerpt from a pre-existing recording even though the length of a particular break was longer than the available sample time.

Hank Shocklee, of Public Enemy's Bomb Squad, often sampled the sounds of pre-existing recordings at the wrong speed. LPs designed to be played at 33⅓ rpm were played at 45 rpm so that a longer excerpt could be sampled. The pitch of the sampled recording was then shifted downwards afterwards. Shocklee described how '[t]he way we stretched time, you lose a little fidelity that way. But back then, who cared about fidelity?' (quoted in Milner 2009, p. 333-334). The designers of the SP-1200 were concerned about what they perceived to be the poor quality sounds of the device. Wedge of E-mu admitted: 'It was okay for a drum machine, but it had cheesy pitch shifting. It got away from the fidelity and quality we aimed for' (quoted in Milner 2009, p. 332). Ignoring fidelity, hip-hop producers were more concerned with the amount of available sample time. Referring to recordings he produced in the late 1980s, RZA of Wu-Tang Clan explained:

> If you lower your sample rate – from 44 kHz to 17-20 kHz – it increases your sample time. So, you get to have longer samples, but with lower resolution. That gives more of a grindy sound, because the sound breaks up. If you lower the sample rate, that means you're missing some of the frequency of the sample. Years later, I heard people call it 'lo-fi', but I just thought it sounded more ghetto and it let me use more sounds.
>
> (2005, p. 197)

Not only did sampling drum machines like the SP-1200 and other digital samplers and sampling keyboards have more sample time and memory, but fixes were found so that longer excerpts of pre-recorded sound could be sampled. 'Inferior' levels of fidelity that were the result of using low sample rates became part of a specific hip-hop aesthetic.

The success of the SP-1200 was a surprise to its designers at E-mu who could not understand why users embraced an instrument with fidelity levels they considered unsatisfactory. For hip-hop producers, Hank Shocklee and Pete Rock, the technological limitations of the instrument contributed to its unique 'feel' and the SP-1200 became their sampling instrument of choice. Speaking in 2008, Pete Rock explained:

> I've done everything you've ever heard from me on the SP-1200 except for this new album where I'm using the [Akai] MPC 2000 XL, and the SP-1200. In the beginning I was working with the SP-12 and the [Roland TR-]909. I liked the feel of the SP-12, and once the SP-1200 came out I basically just fell into it.
>
> (quoted in Mason 2008, p. 57)

For Shocklee, problems with features such as quantisation became a positive: 'It quantised sound very abruptly. It gave the SP-1200 its soul' (quoted in Milner 2009, p. 334). Despite this positive relationship between the instrument and (these) users, the SP-1200 was expensive to produce. It was difficult for E-mu to find parts and they eventually discontinued it in 1990 (Keeble 2002). Alpert explained:

> We'd have to hunt around on the after-market and go through discontinued-parts brokers to get the pieces to keep building them. But every time we announced we were discontinuing it, there would be this hue and cry, with people offering twice as much as [the recommended] retail [price] for them.
>
> (quoted in Milner 2009, p. 332)

As well as manufacturing problems, the designers at E-mu were unhappy that the SP-1200 was used in hip-hop because of controversies around the genre. Scott Wedge explained:

> We tried to stuff it back in the closet. Rap had a bad [reputation]. Politically, it was really ugly stuff. We kind of pulled [the SP-1200] out of retirement, but then we learned that what it was being used for was this rap music, we went, 'Well, let's discontinue it, maybe that'll stop it'.
>
> (quoted in Milner 2009, p. 332)

This is not an example of a sampling instrument being used in ways that were unforeseen by its designers: the manual for the SP-1200 encouraged users to sample sounds from pre-existing recordings on tape, CD, and vinyl. It is, however, an example of the instrument being used in ways that were perceived to conflict with the countercultural image and hippie values that E-mu had tried to cultivate as a company since the 1970s.[29]

While the SP-1200 was one of E-mu's commercially successful products, the company continued to experience technological difficulties with other sample-based instruments. In June 1987 at NAMM in Chicago, E-mu launched the Emulator Three Digital Sound Production System. Rossum explained:

> [W]e wanted to give our E2 [Emulator II] customers the additional features they had requested: true 16-bit fidelity; stereo samples; more channels; more, user-installable, memory; and a richer feature set. While we accomplished that, there was one fatal problem: reliability.
>
> (quoted in Grandl 2015b)

At a cost of $15,000 [US] or £8,000 plus VAT in the UK (Wiffen 1988), the Emulator III sampling keyboard offered users 40MB of hard disk space and 4MB of RAM, which could be expanded to 8MB. A sample rate of 44.1 kHz provided CD-levels of sound quality and 47.2 seconds of sample time. However, owners started to report problems caused by a defect in the SIMM (Single In-line Memory Module) sockets used for the sample memory. Owners returned keyboards to E-mu who issued a recall and replaced all products. Rossum adds: '[T]he EIII's reputation had been ruined. It took us years to recover' (*ibid.*). Again, E-mu had to make staff redundant and was on the verge of bankruptcy. A licensing deal with Matsushita/Technics provided the company with enough capital to release another product called Proteus/1 at Winter NAMM in 1989.[30] The following year, they launched the EMax II all-digital sampler and, in 1993, a rack-based sampler, the Emulator IIIx. However, financial problems persisted and after entering discussions with the Singapore-based company, Creative, an acquisition deal was signed in March 1993. Rossum remained with the company, Wedge stepped down as CEO, and Alpert joined Akai as a marketing consultant in 1994. Rossum admits that E-mu 'didn't worry much about the competitive entries from Casio, Akai, and Roland' (quoted in Grandl 2015b). Like Fairlight Instruments, the company's failure to focus on selling less expensive sampling instruments to a larger number of users was a major reason for its financial instability.[31]

Notes

1 This was changed to E-mu Systems in 1972 when the company became a corporation and Californian law stated names must only use letters from the Roman alphabet.
2 Rossum explained: 'In the early days, E-mu was just located wherever I lived – in my dorm room at UCSC, the house we rented at 625 Water Street during the summer of 1971, and spare bedrooms at other houses' (quoted in Grandl 2015a).
3 In email correspondence, Rossum confirmed: 'The first system we sold to Ed Rudnick (soon to be our employee) in April 1973 [was] an absurdly low price for its content, around $4,000. Bob Moog told us our prices were way too low, we should more than double them, and we did' (Rossum 2018).
4 With an engineer called Ron Dow, Rossum developed Solid State Micro (SSM) integrated circuits that were used on the Prophet-5 and other analogue synthesizers. For more on E-mu and Rossum's relationship with Oberheim Electronics and Sequential Circuits, see Lee 1981.
5 FIFO is an acronym for First, In, First, Out, a way of organising data in a queue-like structure.
6 Hardcastle explained: 'I got hold of an Emulator. It was one of the first samplers and I had hired it for the day for something else. I just started mucking about and recorded me saying, 'Nineteen'. I was fooling around and doing something rhythmic on one key with that sample, going, 'N-N-N-N' and thought this would be a good idea' (quoted in Cunningham 1998, p. 316).
7 Marley Marl was an assistant to Arthur Baker at Unique Recording Studios in New York: 'I was an intern, and I'd just hang around sessions for the Force MDs, [Afrika] Bambaataa, Jazzy Jay and 'em' (quoted in Nelson 1991, p. 38).

8 Mark Katz defines a break as 'a brief percussion solo, typically found towards the end of a funk song, though it may show up anywhere in a song and, really, anywhere in music' (2012, p. 14). In the early years of hip-hop, DJs and producers used drum breaks from recordings by soul and funk artists such as James Brown and The Jimmy Castor Bunch before breaks began to be sourced from a wider range of musics and not only percussion or rhythm sounds. Joseph Schloss explains: 'Today, the term 'break' refers to *any* segment of music (usually four measures or less) that could be sampled and repeated. . .In contemporary terms, a break is any expanse of music that is *thought of as a break* by a producer [original emphasis]' (2014, p. 36).

9 The Captain Rock record Marley Marl was remixing was either 'Cosmic Blast' or 'Capt. Rock to the Future Shock', which were both released by NIA Records in 1984.

10 The spec sheet section of *Keyboard* magazine in August 1981 contains information and prices for Oberheim's DSX sequencer and DMX drum machine. The cost of the DSX was $1,700 [US].

11 Kurtis Blow used the drum machine to create a particular sound: 'The [Roland TR-]808 is great because of the bass drum. You can detune it and get this low-frequency hum. It's a car speaker destroyer. That's what we try to do as rap producers – break car speakers and house speakers and boom boxes. And the [Roland TR-]808 does it. . .' (quoted in Dery and Doerschuk 1988, p. 34).

12 For a review of the Drumulator, see Aikin 1983.

13 The instructions in E-mu's manual for the Drumulator reflected their countercultural attitudes and playful approach to business: 'You can use this manual to line bird cages, as kindling when starting a fire in the fireplace, as the raw material for creating paper gliders, or most importantly, as a guide to help you get the most out of the Drumulator' (E-mu 1983b, p. 1).

14 Gotcher recalled: 'I was doing a lot of recording at the time and didn't really like the sounds I was getting from my drum machine. Evan figured out how the sounds were mapped on the ROMs, E-mu helped a little, and we produced about five alternate sets of sounds. We played them to the folks down at E-mu and they thought that our sounds were much better than their existing ones' (quoted in White 1995, p. 36).

15 After its release in 1984, Brooks and Gotcher purchased an Apple Macintosh computer to edit samples on screen and developed a program called Sound Designer that was used with the Emulator II. Brooks described it as 'an editing environment that allowed you to view, edit, and process sounds' (quoted in Milner 2009, p. 335). They demonstrated the follow-up, Sound Tools, at NAMM in January 1989 as a 'tapeless recording studio'. The company was renamed Digidesign and, in 1991, they released the first version of Pro Tools.

16 According to Rossum, sales of the Drumulator slowed after Sequential Circuits released their first digital drum machine, Drumtraks, in 1984 (Abildgaard 2012). It was part of the Traks Music System, which also included a multitimbral synthesizer with built-in digital recorder called Six-Trak and a Model 64 MIDI Sequencer that connected to a Commodore 64 computer. The suggested retail price of Drumtraks was $1,295 [US] (Sequential Circuits 1984a) and £950 (Sequential Circuits 1984b) in the UK.

17 As well as the Fairlight CMI, Syco Systems distributed products by E-mu, Kurzweil, PPG, Quantec, Linn, and Yamaha. Syco paid E-mu an advance of $100,000 in return for the first forty Emulator II keyboards being shipped directly to the UK for sale to their customers (Cole 2015).

18 The Emulator II was bought by a long list of well-known pop and rock musicians including Stevie Wonder, Herbie Hancock, Paul McCartney, David

Bowie, Vangelis, Jean-Michel Jarre, and Stevie Nicks as well as groups like Genesis, Yes, Depeche Mode, Talking Heads, Orchestral Manoeuvres in the Dark (OMD), and Ultravox (Rossum, quoted in Grandl 2015b).

19 The use of what the engineers at E-mu called the 'sigma-delta encoding scheme' or '8-bit companded' increased the fidelity levels of the Emulator II to 12-bit or higher. It also included features such as analogue synthesizer filters, envelope generators called VCAs (Voltage Controlled Amplifiers), and an SMPTE-based multitrack sequencer. The use of Small Computer System Interface (SCSI) meant the keyboard could be connected to an Apple Macintosh II computer and used with Digidrums' Sound Designer software.

20 The Linn 9000 Integrated Digital Drums/MIDI Keyboard Recorder was launched in 1984 with a price tag of £4,500 plus VAT and distributed in the UK by Syco Systems (Wiffen 1985). It was designed for drum sounds to be loaded from cassette or floppy disk and to digitally sample external sounds, but this was not possible until software updates were made available a year later.

21 Ensoniq introduced the Mirage sampling keyboard in December 1984 for $1,695 [US]. For more on the company and the instrument, see Hastings 1986a, Hastings 1986b, and Anderton 1988b.

22 He writes that the SP-12 was 'well beyond the budget of most inner-city teens' (2014, p. 30). See chapter two of Rose 1994 for a detailed discussion of the urban context of hip-hop production. These include social policies and events in the 1960s-1970s that led to New York and the South Bronx being defined in the US as 'national symbols of ruin and isolation' (p. 33).

23 Paul Théberge writes: '[T]he 'democratisation' that Bob Moog refers to is related to little more than the breaking of the early price barriers that had kept the synthesizer from becoming a broad-based consumer item until the 1980s' (1997, p. 73).

24 Ced-Gee or Cedric Miller was also the producer of the hip-hop group, Ultramagnetic MCs.

25 Along with 'Amen, Brother' (1969) by The Winstons, 'Funky Drummer' (1970) by James Brown is one of the most sampled recordings in the history of popular music. Artists such as N.W.A. and Public Enemy have sampled the drum break by Stubblefield. For more on 'Funky Drummer' and its use in hip-hop, see Stewart 2000, Danielsen 2006, Danielsen 2010, and Scannell 2012. On the sampling by hip-hop and jungle producers of Gregory Coleman's drum break in 'Amen, Brother', more commonly known as 'the Amen break', see Butler 2006, Whelan 2009, and Nate Harrison's audio installation/short film, 'Can I Get An Amen?' (2004).

26 Engineer George Drakoulias lists the technologies used in the making of *Licensed to Ill* as Oberheim DMX and Roland TR-808 drum machines, digital reverberation devices produced by AMS, and magnetic tape. The album was recorded in Chung King Studios, New York, which was 'a 16-track, 2-inch, analog studio. . . .[T]he songs were handmade with no automation and the records were really simple: a drum machine, a rapper yelling, and a DJ scratching. It was very simple and basic' (quoted in Brown 2009, p. 45).

27 Marley Marl remixed two of the tracks on *Paid in Full* ('My Melody' and 'Eric B. is President'). As referred to earlier, 'Eric B is President' uses/samples 'Impeach the President'.

28 Oral histories of hip-hop are useful for drawing attention to the role of hidden actors like Adams and other engineers who were responsible for programming drum machines and sampling instruments in recording studios. Others include Charlie Marotta (EPMD), Shane Faber (A Tribe Called Quest), Schlomo (DJ Premier), and Steve Ett (Public Enemy).

29 In 1994, E-mu relaunched the SP-1200 with a marketing campaign targeted specifically at hip-hop producers: 'Notice how the major rap and hip-hop producers always seem to come up with those 'signature' grooves that rattle your bones? Check out the SP-1200 sampling drum machine from E-mu – those grooves start right here. That's right, the machine that you thought was gone is back by popular demand and as BAD as ever' (E-mu 1994).

30 Before its release, the in-house product name for the Proteus was the Plug: '[W]e had a big hole in our revenue due to the EIII [Emulator III] reliability problems and needed something to plug into the product map as soon as we could' (Rossum, quoted in Grandl 2015c).

31 In 2011, Rossum left E-mu and Creative to work on mobile phone voice technologies for a Silicon Valley-based company called Audience. In 2015, he was re-united with Marco Alpert when they started a company called Rossum Electro-Music.

Interlude

Methodologies

The cultural history of hip-hop presented in the previous chapter was an attempt to shine a sceptical light on some of the claims and orthodoxies in the academic literature on hip-hop and technology. Rather than accepting arguments about the origin points of particular musical practices or the democratisation of music technologies, my aim was to introduce a more nuanced historical account of how technologies were used in the development of hip-hop. This tends to be in more complex and contingent ways than myths that develop around music making suggest. Synthesizers and samplers did not simply replace session musicians in recording studios in the early 1980s; these musicians often learned how to use synthesizers, drum machines, and samplers. Assumptions about the 'affordability' of sampling technologies and the widespread use of instruments like E-mu's SP-12 are contradicted by empirical evidence. Existing technological and musical practices were not immediately displaced; electric instruments and analogue technologies (magnetic tape, turntables) continued to be used alongside new sampling technologies.

Rather than an abrupt or revolutionary shift from analogue to digital, the introduction and use of sampling technologies was part of a longer and more gradual process of socio-musical change. As Paul Théberge argues, the 'digitalisation [of music] has been. . .a relatively long, transformative process of economic, technological, social, and cultural change that has taken place over a half-century or more' (2015, p. 329). It may be problematic, however, to talk about the digitalisation of music. This is partly an ontological distinction, which recognises music as a social process and not a thing that can be digitised (Bohlman 1999).[1] It may be more appropriate to refer to the entanglement of analogue *and* digital technologies as they continue to co-exist in the production and consumption of music.

As well as myths relating to democratisation and digitalisation, some hip-hop journalists and academics have presented a nostalgic argument about the 'golden age of sampling', which was disturbed by the intervention of copyright law and court cases.[2] Jimmy Castor's action against the Beastie Boys and Rick Rubin in 1987 is assumed to have begun a series of lawsuits in the US over the use of sampling technologies.[3] In 1991, The Turtles brought a case against De La Soul for the use of an excerpt from the song, 'You Showed Me', on *Three Feet High and Rising* (1989).[4] These were settled out of court before US District Court judge Kevin Thomas Duffy considered Biz Markie's use of Gilbert O'Sullivan's song 'Alone Again (Naturally)' (*Grand Upright v. Warner* 1991) and decided that any extracts (or samples) from pre-existing recordings should be cleared with the copyright owner.[5] There had been disputes over the ownership of copyright as soon as hip-hop was recorded in the late 1970s and early 1980s.[6] The use of pre-existing sounds in hip-hop did not begin with samplers: as outlined in the previous chapter, the Beastie Boys used magnetic tape to loop a Led Zeppelin break on *Licensed to Ill*. The clearing of samples became an expensive and bureaucratic exercise, but sampling did not slow or stop.[7]

A loop-based approach to using sampling technologies continued to be important to the aesthetic of hip-hop throughout the 1990s. Producers like The RZA, DJ Premier, and Dr Dre were forced to be more inventive in their hunt for obscure sample sources, which were manipulated in ever more creative ways to avoid legal detection. What became more difficult after the case against Biz Markie was using a large number of samples in one recording. Kembrew McLeod explains how copyright law impacted on the production of hip-hop and describes Public Enemy's *Apocalypse 91...The Enemy Strikes Back* (1991): 'Gone were the manic collages that distinguished their previous two albums, where they fused dozens of fragments to create a single song' (2007, p. 68). As well as the legal obligation to clear samples, however, there were other reasons for the changes in their sound. The production team, The Bomb Squad, who had worked on the previous two albums did not produce *Apocalypse 91...* because of anti-Semitic controversies and arguments over album credits.[8] What was different in the mid-to-late 1980s was hip-hop's growing popularity and commercial success. The old recording industry adage that 'where's there's a hit, there's a writ' meant claimants could seek large sums of money if a song (and recording) was used without the permission of the copyright owner(s). What high-profile court cases over copyright infringement in the late 1980s and early 1990s made more difficult was a *particular* approach to using samples in (hip-hop) music.[9]

Some copyright scholars have accepted the argument about the effect of legal action in the US on the musical practices of hip-hop, and the use of sampling technologies more generally. Friedemann Kawohl and Martin Kretschmer write: 'Following numerous restrictive court decisions (culminating in the US case *Bridgeport v. Dimension* 2005), the aesthetics of sampling changed quite dramatically' (2009, p. 220).[10] Other commentators and academics were more dramatic and referred to the death of sampling.[11] Despite the potential threat of litigation, however, users of sampling technologies have continued to appropriate pre-existing materials in hip-hop as well as remixes and mash-ups. As producer Danger Mouse discovered after illegally releasing *The Grey Album*, copyright infringement may be as likely to lead to an increase in career opportunities as the threat of bankruptcy.[12] In countries other than the US and in genres of popular music where appropriation is less important, sampling practices have been largely unaffected by these court cases. This will be shown in the second half of the book with chapters that highlight a diverse range of practices relating to the contemporary use of sampling technologies. I broached the issue of copyright with the musicians and users who were interviewed for its four case studies – Matthew Herbert pointed to a possible scenario where copyright law might be used by organisations and corporations to claim ownership of sounds that are part of the public sphere and Kenny Anderson (aka King Creosote) explained how he prefers to sample pre-existing recordings that are out of copyright – but none complained that it placed restrictions on their creative practices.[13]

Research Methodology

The four case studies in this book explore the use of sampling technologies by musicians and producers operating in and across a variety of different genres at the start of the twenty-first century. In the first case study, I trace the technique of microsampling to the work of two producers (Marc Leclair aka Akufen and Todd Edwards) who are allied with two different sub-genres of dance music – microhouse and UK garage. In the second case study, I focus on Found, a pop group and art collective who are based in Edinburgh and influenced by folk, rock, hip-hop, and electronic musics. I examine their use of sampling/sequencing instruments like Akai's MPC2000 to see if they continue an art school tradition of adopting experimental approaches to pop music using new technologies. The subject of the third case study is a leading member of The Fence Collective in Fife, Kenny Anderson, who has been associated with new styles of folk but rejects the term when it is applied to his own music. He inserts samples into his home-studio recordings by using a guitar pedal to digitally reproduce and loop sounds from a variety of random sources. The final case study is about Matthew Herbert, who is treated by critics and fans as an *auteur* of sampling and who moves between the worlds of dance, jazz, and art musics. His case study is situated differently to the others because he opposes and rejects many forms of popular culture, a position that means he is more closely aligned with the musical worlds of field recording and sound art than those of popular music.

In the first half of the book, I used a combination of archival research and interviews to understand the shifting historical practices of digital sampling in the late 1970s and 1980s. The chapters that follow are based on empirical evidence gained from interviews with users about more recent technological practices. My focus on microsampling in the next chapter is to begin with a non-loop-based approach to the use of sampling technologies and might be described as the 'pilot case study' (Yin 2009). I conducted email interviews with Leclair and Edwards, who are both based in North America. These yielded less data than the focused interviews with other users, which lasted between an hour and two hours and form the basis of the remaining three case studies.

I chose Found because sampling/sequencing technologies and hardware devices, like the MPC 2000 and laptops, were central to both their live performances and their recordings. I was interested in Anderson's work because he was using samples from pre-existing recordings in a genre of popular music – folk – that is not usually associated with sampling. It was also important to focus on users considered influential in the field of sampling and I carried out a face-to-face interview with Herbert. His visit to Edinburgh as part of a Hackathon gave me an opportunity to ask him some questions about sampling in an arts venue called Summerhall. Formerly a School of Veterinary Studies, we searched for a suitable space and found a large laboratory-like room with uncomfortable stools. Opting instead to record the interview in an outside bar area, we were distracted by background noise and light rain that threatened to disrupt the recording by forcing us to move indoors. One of the themes in this book is the role of accidents in the design of musical instruments and also how users of music technologies deal with mishaps that occur during the recording process. The interviews for these case studies were similarly shaped by 'affordances' and slips – the battery time of ageing minidisc recorders, not being able to use technologies confidently – that had an impact on the social practices of gathering research data.

One of my arguments in the book is that sampling technologies are best understood within locally-situated practices of music making, the changing contexts of recording studios (mixing desks, laptops, and software samplers), and the histories of recording and sound reproduction technologies. The use of the term 'sampling' and its definition by users of music technologies is also shaped by changing musical practices. One of the members of Found, Tommy Perman, explained that his definition of sampling had changed along with the shifting nature of his musical collaborations: '[I]t started off for me as stealing bits of other people's music and now it's definitely become a way of manipulating a sound' (2008). Anderson was happy to accept the definition supplied by detractors of sampling that it is a form of theft while Herbert was willing to admit he did not have an adequate definition because sampling technologies and practices are continually changing. Despite the lack of a definitive answer, the case studies are based around diverse approaches to the question 'What is sampling?'. This led to further subsidiary questions: What does sampling mean to these musicians and producers? Why do they define sampling this way? What does it mean to sample music? What does it mean to use samples in music? What technologies are they using to sample? The process of gathering data for the case studies was influenced by Timothy Taylor's question: '[W]hat are these social actors doing in this time and place, and why?' (2001, p. 37) For the purposes of writing and structuring these chapters, I rephrased it slightly: How are these social actors using digital sampling technologies and why?

Notes

1 See chapter six of Sterne 2012b where he defines music as 'a bundle of affordances, thus borrowing some of the process language and some of the thing language' (p. 189). Similarly, Nick Prior employs the concept of 'assemblage' to 'move us away from idealist notions of music as fixed in the score, or in the mind of the creative genius, [or in the sound recording,] towards a more radically relational approach that takes into account the chains of associations between humans, institutions, technologies, texts, tools, instruments, works, and so on' (2017, p. 6).

2 McLeod and DiCola describe the golden age of sampling as 'a moment in time in the late 1980s and early 1990s when artists had more freedom to create sample-based music' (2011, pp. 5–6).

3 Beastie Boys and Def Jam were sued over the use of Castor's 'The Return of Leroy (Part One)' (1977) on 'Hold it Now, Hit it' from *Licensed to Ill*. For more on the Manhattan federal district case, *Castor v Rubin*, 87 Civ. 6159, which was settled out of court, see Marcus 1991.

4 In the UK, a series of disputes relating to the use of sampling technologies to reproduce sounds from pre-existing recordings included JAMS v Abba 1987, SAW v M/A/R/R/S 1987, and Hyperion Records v The Beloved 1991. See Sutcliffe 1987, Beadle 1993, and Frith 1993.

5 For more on *Grand Upright Music, Ltd. v. Warner Bros. Records*, 780 F. Supp. 182 (S.D.N.Y. 1991), see Falstrom 1994.

6 The release of records including Sugarhill Gang's 'Rapper's Delight', Grandmaster Flash and Melle Mel's 'White Lines (Don't Do It)', Malcolm X's 'No Sell Out', and Afrika Bambaataa's 'Planet Rock' resulted in disputes and lawsuits. Some of these were settled out of court, some involved threats of violence. See McLeod and DiCola 2011 and Nile Rodgers' (2011) autobiography for examples of intimidation after legal cases brought against Sugarhill Records.

7 Referring to the consequences of the *Grand v. Warner* case, Siva Vaidhyanathan described how 'rap music since 1991 has been marked by a severe decrease in the amount of sampling' (2001, p. 143). According to Joanna Demers, though, this is to 'oversimplify a complicated situation. . .After 1991, sampling persisted among artists on the extremes: those who were either rich enough to afford licensing or obscure enough to be able to risk illegal sampling' (2006, p. 97).

8 *Apocalypse 91. . .* was produced by The Imperial Grand Ministers of Funk; The Bomb Squad were Executive Producers. See Myrie 2008 for an authorised biography of Public Enemy and material on its internal disputes. For Chuck D and Hank Shocklee's views on how their approach to sampling changed as a result of court cases related to use of digital sampling, see McLeod 2004.

9　For more on the relationship between digital sampling, copyright law, and practices of appropriation across a range of cultural forms such as collage, montage, and remix, see Bourriaud 2002, Miller 2008, Boon 2010, McLeod and Kuenzli 2011, Laderman and Westrup 2014.

10　For more on *Bridgeport Music, Inc. v. Dimension Films*, 410 F.3d 792 (6th Cir. 2005) and related cases, see Théberge 2004, Schietinger 2005, Mueller 2006, and Morey 2012.

11　See Kemp 1992, Marshall 2006, and Morey 2007 for further discussion of this issue.

12　Danger Mouse, aka Brian Burton, was instructed to cease and desist distribution of *The Grey Album* (2004), which contained music from The Beatles' untitled 1968 album – more commonly known as *The White Album* – and rap *a cappellas* from Jay-Z's *The Black Album* (2003). He went on to produce albums for Beck, The Black Keys, and Gorillaz, and formed Gnarls Barkley. For further discussion of this case and other mash-ups, see Brøvig-Hanssen and Harkins 2012.

13　As Demers argues: '[S]tudying the effects of IP law on music by looking only at major label talent is to ignore the vast majority of musicians who do not appropriate from famous artists or who exploit loopholes in copyright law to their artistic and financial advantage' (2006, pp. 113–114).

PART II

USERS

FOUR

Microsampling

Akufen and Todd Edwards

Introduction

This chapter investigates a style of sampling known as microsampling and is situated within two sub-genres of electronic dance music (EDM): microhouse and UK garage. I focus on the practices of two producers who, instead of using sampling instruments to loop drum patterns from pre-existing recordings, design rhythms and melodies at the micro level by manipulating recorded sound from various sources. The term microsampling can be traced to the music and ideas of microhouse producer Marc Leclair (aka Akufen) and is also relevant to the music of Todd Edwards and its influence on the sound of UK garage. Using data from email interviews with Leclair and Edwards as well as secondary sources, this chapter sets out to define microsampling and examine the technologies used to make music with microsamples; for example, both Leclair and Edwards use hardware samplers and software samplers. This case study focuses on two users for whom the sampler is their main instrument and is an example of how the development and use of sampling instruments since the 1990s has shaped musical practices and creative possibilities around the digital manipulation of recorded sound.

Microhouse, Microsampling

With an uneasy relationship to the American roots of house, producers of microhouse like Jan Jelinek, Thomas Brinkmann, and Isolée were mainly based in German cities: Berlin, Cologne, Frankfurt.[1] Marc Leclair, though, is a Canadian from Montreal and his pseudonym, Akufen, is a verbal play on the French word for tinnitus (acouphène). He began using the term microsampling in 2001 after developing a specific approach to using digital sampling technologies in the mid-1990s. He defines it as

> . . .borrowing a very short amount of sound matter to a
> point where it will not be recognisable, partly for decency
> and respect for the work of others. More specifically, the
> main idea behind my work is to recycle sound particles into
> a larger organism, which is, in this case, a musical piece.
> You can listen or look at it at a larger scale or you can dive
> deep into its complex structure and dissect it. Sampling is
> very three dimensional because each source is unknown
> to the other. Each sound has its own character and grain.
>
> (Leclair 2008a)

One of the precursors to Leclair's idea of microsampling is John Oswald's
CD *Plexure* (1993), which contains over 4,000 small samples or 'electroquotes' of more than 1,000 pop songs. Chris Cutler writes: '[There are] so
many tiny cuts and samples on it that. . .their identities [are] impossible
to register by listening' (1994, p. 16). According to Cutler, a 'macrosam-
ple' was Oswald's term for the 'capture and re-use of entire recordings
as opposed to extracts or snatches of existing recordings' (2008). The
method is most obvious on the copyright-infringing CD, *Plunderphonic*
(1989).[2] Despite similarities in their practice and a shared Canadian background, Leclair rejects the idea that Oswald's use of 'electroquotes' or
macrosamples inspired his own approach:

> Absolutely not, even though I believe his work is
> very relevant, interesting, and unique. John Oswald's
> Plunderphonics projects were more political, like
> Negativland's work. My approach is more aesthetic and
> artistic. Their sampling is intentionally obvious. They
> want to create a reaction by seeking the most straightfor-
> ward sample references from the popular catalogue. John
> Oswald speaks about 'quoting other music', which I think
> is very well phrased. I do everything but quoting other
> music. I want people to forget about where it comes from.
> I wish the elements in my work to gain a new life and
> become part of a new piece of music as if it was for the
> first time ever. This fraction of a second is now a note, a
> sound recontextualised.
>
> (Leclair 2008a)

Continuing the process R. Murray Schafer refers to as schizophonia,
which began with sound recording and involves 'the splitting of sounds
from their original contexts' (1977, p. 88), Leclair detaches samples from
their sources so they are unrecognisable and, measurable in seconds or
milliseconds, more accurately referred to as microsamples.

Leclair's use of sampling technologies began with hardware samplers and one of the 'affordable' sampling keyboards developed in the 1980s: an Ensoniq Mirage. He described how it 'changed my whole perception of making music. And it changed the face of electronic music forever' (Leclair 2008a). Here, the discourse of technological determinism is reproduced, and the role of users is elided. Leclair went on to use

> . . .the phenomenal Casio FZ-1 sampler on which I did most of my homework and training. I loved the raunchiness of its filter and the grain it added to the sound. It was the favourite of artists such as Richard D. James aka The Aphex Twin. I can understand why. It was built to be creative, more than for just recording. Akai were also notable pioneers of sampler development, but they never built a machine that suited my needs. It was more a straightforward digital recorder. It never topped Casio's or Ensoniq's creative possibilities.
>
> (*ibid.*)

For Leclair, his use of sampling technologies is associated with a specific ideology of creativity and authenticity and is also based on an ethical position. It explains his lack of interest in using pre-existing recordings as source material for his own recordings:

> I've been always very respectful of the work of others. A sampler is a powerful and creative instrument, but it has a code of ethics that should be respected. This is my belief. It might not be embraced by everyone and I respect that also. Sadly, it was misused during a certain time in music history to rip-off artists.
>
> (*ibid.*)

While keen to encourage, what he refers to as, the democratisation of the sampler as an instrument, Leclair expresses distaste for the appropriation of pre-existing recordings (or artworks) and explained to me: 'I will never engage in a path of voluntary thievery' (*ibid.*). He considers approaches to the use of sampling technologies that are associated with stealing as *less* creative and showing a lack of respect for other artists' work, even though the permission of copyright owners is obtained when samples of pre-existing recordings are cleared. Rather than celebrating how technologies have been used in ways other than those imagined by their designers, Leclair voices regret about the use of hardware technologies in the 1980s to sample pre-existing sounds as an example of *misuse.*

Having initially used sampling keyboards like Ensoniq's Mirage and Casio's FZ-1, Leclair began to use personal computers (PCs) and software samplers for microsampling.[3] As the amount of RAM available with PCs continued to increase, he was able to store digital recordings lasting hours rather than seconds or minutes:

> When the computer made its first appearance it was like locking a child in a candy store. Sampler users were to that day limited to a very restricted amount of recording time, 2MB, which you could upgrade sometimes, but the cost was obscene. With the PC you could now record hours of sound matter. So, the possibilities became endless. The world became our sound source and, with the infinite possibilities of distorting and altering sound, we were now at the dusk of a new blossoming creative explosion.
>
> (*ibid.*)

Along with enthusiasm for the options available to PC users who wanted to digitally sample everyday sounds, Leclair highlighted some of the limitations of using a mouse, keyboard, and monitor when making music. When I asked if the sampler allowed sound to be sculpted in a similar way to how visual artists work, he replied:

> The difference is the direct contact with the matter. A lump of clay and a knob isn't quite the same. The rotary and redundant movement of a knob or a computer mouse isn't close to the organic movement of a drawing or sculpting hand. With a computer you are a bit limited with the movement that is crucial in sculpting or drawing. Most of the work is done internally by the machine. More and more, though, we will see interactive interfaces where you can hold a pencil and draw the wave of your sound and also have a screen where you can manipulate the sound with your hands. It will eventually resemble sculpting.
>
> (*ibid.*)

Leclair imagines drawing sound waves with technologies that resemble the light-pen used to draw waveforms on the monitor of the Fairlight CMI Series I. He envisages a 'closer' physical relationship between the hands and bodies of users and the production/reproduction of recorded sounds, one that is more active and less constrained by technologies that are viewed as interfering with an 'organic' creative process.

Along with software samplers that could be used to record and edit digital audio, software synthesizers were developed as a way of creating virtual versions of hardware instruments.[4] Leclair is critical of both hardware companies and those who design simulated versions of analogue instruments like Roland's TB-303 Bass Line synthesizer:

> I've never had such a good relationship with the hardware companies. Every now and then there is, of course, a ground-breaking new technology but 90% of the products on the market are pretty much doing the same stuff: emulation of this and that. How many TB-303 emulations have been done? All of them are fairly close but none of them will ever capture the essence of the real thing.
>
> (*ibid.*)

The discourse of authenticity around the use of 'real' sounds also extends to 'real' instruments. While still using acoustic instruments – Leclair told me he plays piano for at least five hours a day – he now mostly uses software samplers and synthesizers, partly because of domestic space restrictions: 'I do have hardware still, but due to space inconvenience I have to limit my studio to my bedroom, which is about the size of my bed' (*ibid.*). He re-introduced analogue synthesizers into his studio set-up but only temporarily:

> Every now and then I do plug in my Doepfer modular [synthesizer] and tweak it but it's more like a child who finds a toy in the bottom of a bin and rediscovers it, until he gets bored again and trades it for something else. I get tired quickly with gear. I still can't believe that some people can buy a device that plays one or two sounds for obscene amounts of money when they have access to the largest sound bank at the tip of their finger: the world. I mean one can argue and speak to me for hours about the purity and fatness of the analogue kick drum of the [Roland] TR-808 but I still think that it is insane to limit ourselves to the gear that multinationals are trying to impose on us. That's why a lot of the electronic music out there sounds so similar and unchallenging.
>
> (*ibid.*)

This argument will re-appear in the case study of Matthew Herbert who expresses a similar frustration with the homogeneity of sampled and pre-recorded sounds. Like Herbert, Leclair's solution is to expand his sonic palette by sampling the sounds of the world. In his case, 'the sounds of everyday life' are mediated by radio broadcasting technologies.

'Everything is Recyclable': Resuscitating and Reviving Radio Waves

Leclair collects microsamples by surfing radio stations and sampling random fragments of sound. These include unidentified songs or white noise from a mistuned analogue signal, which is then recontextualised into new recordings. He explains his methods:

> I sample hours of radio airwaves every morning and dissect fractions or seconds of them to a point where samples aren't recognisable. Then I assemble every bit like a puzzle, or a collage if you prefer. It's a long process and I never know what I'm gonna end up with. My approach is very much inspired by the surrealistic techniques and the French-Canadian automatists, like painter [Jean-Paul] Riopelle and writer [Claude] Gauvreau. I like the error margin and the unexpected factor, which often makes a lot of sense subconsciously, so I have to be very spontaneous in my way of working. I find a lot of essential answers in my music – it's like psychoanalysis.
>
> (quoted in Herrmann 2002)

Inspired by funk as well as Freud, Stevie Wonder is as significant an influence on Leclair's creativity as Steve Reich and Bill Evans. His musical practices have been shaped by 'sampling virtuosos' (Leclair 2008a) including Negativland, Matthew Herbert, and Uwe Schmidt. However, his earliest experiments with the use of sampling technologies were stimulated by the recordings of industrial music and post-punk artists:

> The sampler just came at a moment in my life when I needed this little extra ingredient to spice up my music ideas. I'd say it was a trigger to my ideas. At the time I was very much listening to bands like The Residents, Severed Heads, or Throbbing Gristle and they were all using samplers. I was always wondering where they'd got those abstract, and sometimes Dadaist, soundscapes.
>
> (*ibid.*)

With an academic background in the study of the visual arts, Leclair refers to the canons of modern art and literature to pay homage to the historical roots of those forms of sampling that involve musical borrowing. For him, microsampling is a musical version of collage. Unlike the sample-based 'collages' of hip-hop, though, he does not extract his sound sources from recordings that he owns on vinyl or CD but from radio broadcasts that are transmitting the sounds of pre-existing recordings along with other sounds.

Early recording technologies were designed for the preservation of sound, including human voices that could be listened to after the death of their owner.[5] Leclair seeks immortality for *digitally* recorded sounds while, at the same time, treating them as if they are as recyclable as glass or plastic. As well as explaining how his sonic collages are inspired by the images of surrealists like Andre Breton and the literary cut-ups of William Burroughs, he refers to himself as a photographer of sound. Using sampling technologies is a way of taking sonic photographs and, for him, is about permanence:

> Sampling is like taking pictures. I see myself as a photographer. I take snapshots of sound and immortalise them forever, seconds of unpredictable soundwaves crossing paths at a given time. This will never come back again, and nothing else will ever sound like it. It's unique, like everything else in nature.
>
> *(ibid.)*

With an interest in the sounds of human/non-human environments or what might be called the 'aural public sphere' (Ochoa Gautier 2006), Leclair wants to enact the role of what Arielle Saiber calls an 'acoustic microsurgeon' (2007, p. 1618). He tries to resuscitate and revive radio sounds that would otherwise disappear into the ether:

> I attempt to give new life to dead airwaves caught on the very moment of their short existence. My studio has become a graveyard for those dead frequencies. I take pictures of those dying waves and immortalise them in my software sampler – hours of whatever's lying there. From there I just have to dissect parts, organs that are still usable. Whether a part was a success or a failure, there's always something to recycle in order to give it new life. A fraction of a vocal, of a pad, a glitch, or interference integrated with an advertisement or a song – everything is recyclable.
>
> (Leclair 2001)

As with early sound recording technologies, a tension exists between the idea of using technologies to store sounds indefinitely and the malleability of these sounds, which, in this case, are reorganised and recontextualised so that the source is unknowable.

The opening of the track 'Deck the House' by Akufen makes it difficult to detect a rhythmical pattern among the microsampled snatches of syllables and instrumental sounds that might be guitar strums or saxophone bursts.[6] Their origin is unknown, however, and impossible to locate with any certainty due to the abbreviated nature of the notes. Simon Reynolds describes the effect as 'choppily post-modern and fractured, making me imagine what it might be like to inhabit the scatterbrain of someone who's eighteen and has barely known a world without videogames, an infinity of TV channels, and MP3s' (2003). Leclair uses sampling technologies to construct musical collages and melodies using a juxtaposition of random microsamples that can cause feelings of disorientation. The approach to rhythm, though, is metronomic and different to the skip and swing associated with the drum programming in the recordings of Todd Edwards, which I turn to next. Rather than using microsamples to create a dizzying sensation, Edwards uses them to create chord arrangements that are pleasing both to him and his audience. However, the vocal melodies he constructs may still cause confusion due to their lack of meaning.

Todd Edwards and his Sample Choirs

While Leclair began using 'microsampling' to describe his musical practices at the turn of the century, Edwards was unaware of the word when I interviewed him: 'I was unfamiliar with the term microsampling until I searched for it online. I don't know if it relates to my work. I can say that samples in my tracks can range from blips that are usually unrecognisable to vocal phrases and musical riffs' (2008). Like Leclair, Edwards manipulates sounds so that the source is not recognisable but is more likely to sample and use vocal sounds compared to Leclair's largely instrumental tracks. Although Edwards was unsure whether his musical practices should be described as microsampling, critics have drawn comparisons between his music and the arrangement of microsamples on Akufen's *My Way*.[7] When I interviewed Leclair, he was sensitive to accusations of plagiarism and frustrated with the inability of music journalists to see the subtle differences between his use of microsamples and that of Edwards:

> I like and respect Edwards' work very much, but it's nothing like mine. We're working differently and we're coming from different backgrounds. He mainly uses the cutting technique on vocals, while I use it on the whole song and get my sources from all over the place: radio, TV, movies, field recordings.
>
> (Leclair 2008b)

Leclair and Edwards are part of the broad church of house music but are attached to different denominations. While Leclair has been closely aligned with the European glitch of microhouse, Edwards began his career duplicating disco music rooted in African-American traditions. He then went on to develop a vocal cut-up style, which had a major impact on producers and DJs who were part of the UK garage scene in London.[8]

Within the UK garage scene, the New Jersey-born producer Edwards has been granted special status as the 'godfather of UK garage' and referred to as 'Todd The God' by followers (Read 2001).[9] He started making what he calls 'club music' in 1989 and a few years later began to experiment with the sounds of cut-up samples used by another US-based producer, Marc Kinchen or MK. When I asked about his musical influences and any producers he admired for their use of sampling technologies, he told me:

> It started and ended with MK. . .. He was best known for his dub remix of 'Push The Feeling On' (1992) by Nightcrawlers. It had a haunting bassline with a vocal hook that sounded like something was being sung over and over but, in actuality, the hook wasn't singing anything. They were just syllables that MK pieced together from the original song.
>
> (Edwards 2008)

In one magazine interview, he described this as a moment when 'a light bulb went off in my head' (quoted in Host 2002, p. 20). Edwards realised that short samples of vocal sounds could be arranged to create a melody without forming words that made any literal sense. The idea of meaningless melodies was also inspired by a more unlikely source: the Irish singer, Enya. Her synthesized vocal style is achieved by multi-tracking her voice on hundreds, and sometimes thousands, of tracks to create the sound of a virtual choir or what she calls 'the choir of one' (quoted in Barrett 2008, p. 18). Edwards explained:

> Listening to the works of Enya, I started using vocals as musical instruments. I love the way certain voices sound. It's like the way a flute is different from a clarinet. One singer differs from another. Different sounding voices add different elements to a track, and to a song as well. Voices have different textures. Some are smooth, some are rough, and some are angelic. Certain syllables and words give different effects in a track. I don't try to sample the same things all the time, but I know what is pleasing to my ears.

> There are words that I enjoy hearing for their rhythmic
> qualities. Using vocals as musical elements also made my
> work more identifiable.
>
> (Edwards 2008)

Rather than sampling the sounds of a piano or stringed instruments, and in contrast to the sampling practices of hip-hop producers more likely to search for rhythmic sounds – the perfect beat, break, or bassline – Edwards realised he could sample the voice as a musical instrument. He creates choirs of microsampled voices, which rely on the integration of divergent syllables and sounds to construct new sonic textures, melodies, and chords.[10]

Rather than scanning the radio airwaves, as Leclair does, Edwards finds voices by searching through pre-existing recordings for specific sounds. He explained this while highlighting the central role that sampling technologies play as part of his music making:

> The sampler is the most important instrument I use
> to make my music. I go through records, CDs, and
> MP3 albums searching for musical notes, chords, and
> riffs. They may be instrumental samples, but I prefer
> voices. I build libraries of these and one sample goes on
> each key of the musical keyboard. I look for different
> chord types, primarily major, major 7ths, minor, and
> minor 7ths. They can then be manipulated into chord
> arrangements. Singular sounds and voices help build
> the patterns as well.
>
> (*ibid.*)

Edwards developed an interest in the texture of vocals by artists like Joan Baez, The Carpenters, and Crosby, Stills, and Nash. However, on 'Saved My Life' (1995), a track that had a significant impact on the early UK garage scene, he used his and his father's voice. When asked in an interview about the strangest thing he had ever sampled, he said, '[Laughing]. . .my father's voice! When I did 'Saved My Life', I sang half of the samples myself, but I needed a baritone voice to go in the little choir sound, so I had him come in and sing an 'ooh' for me!' (quoted in Read 1999, p. 55). While recording the voices of family members in this way, Edwards did not face legal problems over copyright ownership. When he does sample sounds from pre-existing recordings, he tends not to clear them, primarily because he is working with short samples where the sources are unrecognisable. On the subject of sample clearance, he told me: 'It would be impossible to clear the samples I use. One track can have up to 100 small samples in it' (2008).[11]

'Saved My Life' begins as a house track before the introduction of a sampled voice repeating a short 'uh' sound along with Hammond organ-sounding stabs.[12] This is followed by the repetition of four lines where the words sung by female voices are difficult to decipher apart from the last phrase, which relays the title of the track. A crash introduces the microsample choir with its angelic sounding 'oohs' and an individual voice enters with much clearer lyrics ('You gave me love, I just can't get enough').[13] The second phrase ('I just can't get enough') is isolated in the second half (the song) and any intended message may have been lost as it crossed the Atlantic; an interpretation of insatiable sexual and/or chemical desire may have been more likely among UK garage audiences. Along with 'Saved My Life', remixes by Todd Edwards are also key to the influence of his vocal cut-up style on the sound of UK garage. His vocal remix of St Germain's 'Alabama Blues' (1995) is a radical reorganisation of the original version, with the insertion of a bridge and chorus with microsamples. It transforms a downbeat story about racial alienation into a vocal expression of overwhelming joy and invites a reading of overcoming adversity.[14] As Edwards explains: 'I improvised the song; if I think something needed a bridge, I'd make one' (quoted in Matos 2007). This complicates the idea of the remix as a deconstruction of the song and shows that the use of sampling technologies did not lead to 'the death of the song' (1990, p. 171) as Simon Reynolds and David Stubbs had hoped.

Using the Ensoniq EPS and Akai S6000: Skip and 'Bumpy Swing' in UK Garage

As well as his use of microsampled voices, the skip and swing in the drum sounds of recordings by Todd Edwards were crucial to their appeal to UK producers and ravers.[15] When I asked how he used sampling technologies to achieve this, Edwards said:

> There were originally two ways this was done. The first was my early attempts of trying to imitate Kenny 'Dope' Gonzalez's drum programming.[16] The second ingredient was the 16-triplet quantising on my Ensoniq EPS sampling keyboard.[17] This was my first sampler. I used it to sequence my tracks as well. It had a really hard 16-triplet quantise. The two concepts gave my drum programming a bumpy swing.
>
> (Edwards 2008)

The quantisation of sounds was first made possible with MIDI and sampling/sequencing technologies like Page R on the Fairlight CMI.[18]

While this is valued by producers working in some genres of music – in Joseph Schloss's study of hip-hop production, one of the interviewees criticises RZA of Wu-Tang Clan for his 'sloppy' beats and failure to use quantisation (2014, p. 141) – the programming of 'perfectly' timed rhythms made possible by using digital technologies is what Edwards decided to try and avoid:

> I used to quantise a lot. When I upgraded my sequencer from the Ensoniq EPS to an actual computer, the software that I used did not quantise the same way. I started doing the 'skipping beat'-style patterns by ear. That was followed by programming the musical elements by ear as well. I think it gives my tracks a more organic feel. . .less robotic.
>
> (*ibid.*)

Edwards' use of the Ensoniq EPS sampling keyboard to sequence sounds with swing is what made them appealing to the ears, hips, and feet of UK garage producers and ravers. His upgrade to sequencing software resulted in him programming sounds without the use of quantisation. Like Leclair, Edwards values the 'organic' and employs digital technologies to retain the 'imperfections' associated with the use of 'real' instruments.[19]

When I interviewed him, Edwards explained that the digital technologies he used as part of his sampling practices still included hardware samplers. Along with switching to software to sequence his sounds, he had also moved from using the Ensoniq EPS sampling keyboard to a rack-based sampler, Akai's S6000.[20] Rather than a narrative of technological progress, Edwards had to solve new problems that this introduced:

> I still use an Akai S6000. Changing equipment has its pros and cons. It allows for growth and the ability to work faster. It can allow you to explore new ideas and add new elements to the creative process. My original Ensoniq EPS had, I think, 30 seconds of sampling time. The simplicity of it challenged me to use what I had to its full potential, improvising and performing tricks to cover up the lack of equipment that I needed at the time. For example, instead of an echo or a delay, I looped a sample, and set it to fade slowly as it looped. Also, the sound quality of the Ensoniq EPS, or lack of quality, became as much an element of the music as any other piece of equipment. What makes music recorded before the 1980s

so enjoyable is the imperfection of it. It was human, warm, crackly, hissing, muffled. Most of my sampling is from works created before the 1980s.

(*ibid.*)

As a user of digital sampling technologies, Edwards expresses a nostalgia for *non-digital* sounds. In the same way that judgements relating to the fidelity and quality of sounds shifted with the development of new sampling instruments during the 1980s, so, too, did perceptions about the amount of sample time. Edwards now views 30 seconds of sample time as restrictive. Working with these 'affordances' resulted in creative fixes.

As with Leclair's use of software samplers, the Akai S6000 gave users more sample time. For Edwards, this led to problems that resulted from having *too* many options:

> Switching to the Akai S6000 gave me a greater amount of sample time, 30 minutes instead of 30 seconds. But it doesn't sound the same. It's cleaner. I was able to broaden the scope of what I could do but it becomes overwhelming at times. There are more samples to choose from. More can be added to one track. I developed a tendency to be extremely complicated in the sample arrangements and, at the same time, fell victim to becoming formulaic and having trouble creating outside the box.
>
> (*ibid.*)

To explain the difference between using the Ensoniq EPS sampling keyboard and an Akai S6000 digital sampler, Edwards drew an analogy with the process of painting. Unlike Leclair, he does not have a background in the study of visual art. However, Edwards had mentioned in an interview that having a bank of samples is 'almost like having your paints ready to paint on the canvas, instead of mixing them as you're painting' (quoted in Matos 2007). I asked him the same question I had asked Leclair about whether sampling technologies allow sound to be sculpted in a similar way to how artists paint:

> How I compose now is closer to having a blank canvas with an array of paints ready to use in front of me. Composing with a sampler that only gave me 30 seconds of sampling time would be like being a painter that has to look through a large number of boxes filled with old paints. The painter has to go through the paint containers one by one. Most of them are dried out. He

> then finds one, and it has a little yellow in it. . .enough
> for one brush stroke. So, he brushes a little yellow
> on the canvas. What did he paint? He doesn't know
> yet. It's just a brush stroke. He will search and find a
> few more colours he likes and the painting develops.
> A couple of weeks later the painting is finished. Of
> course, the painter never got a 'reboot error' before he
> finished saving his work, and then had to start all over
> again on the painting!???.
>
> (Edwards 2008)

The contingencies of musical practices are mediated by the unpredict-abilities of the non-human. Users matter but so do technologies and their bugs and glitches. Where Leclair positions himself in an artistic field of production by referring to his many influences, Edwards shows less interest in art history and is more likely to reflect on his marginal position in US house and garage scenes. Compared to devices with less sample time and smaller sample libraries, he has more options available to him with the Akai S6000 and is able to 'paint with sound' using a greater range of colours and shades. And yet the introduction of newer digital sampling technologies into his workflow involves negotiating more choices, new challenges, and a nostalgia for older technologies.

Notes

1 In *The Wire* in July 2001, journalist Philip Sherburne wrote about a new sub-genre of house music called microhouse in which 'percussive elements – the thumping bass drum, ticking hi-hat, etc – have been replaced by tics and pops and compressed bits of static and hiss' (p. 22).

2 For more on Oswald's Plunderphonics, see Oswald 1986 and 1988, Igma 1990a and 1990b, Cutler 1994, Holm-Hudson 1996 and 1997, Steenhuisen 2009, and Sanden 2012.

3 Early software samplers like Nemesys Gigasampler v1.0 (1998) were designed to overcome the memory limitations of hardware samplers by streaming digital audio direct from the hard drive of a computer. Available for PCs at a cost of £599, Gigasampler required a minimum of 2GB and could be used with up to 18GB of hard drive space (Walker 1998). At this time, the maximum RAM capacity of many hardware samplers was between 8MB and 32MB.

4 Some of the first software synthesizers included Reality, which was developed by Seer Systems and released in January 1997. The President and Head Engineer of the company was Dave Smith, formerly of Sequential Circuits. In the same month, Propellerhead Software introduced ReBirth with simulated versions of Roland's TB-303 and TR-808 instruments. For more on software synthesizers (usually referred to as softsynths), see Ingram 2009, Vail 2014, and Holmes 2016.

5 Jonathan Sterne writes: 'If there was a defining figure in early accounts of sound recording, it was the possibility of preserving the voice beyond the death of the speaker. If there was a defining characteristic of those first recording devices

and uses to which they were put, it was the ephemerality of sound recordings' (2003, p. 287). Early recordings were often unplayable on more than one device and later forms of musical storage, such as shellac, were fragile and unreliable.

6 'Deck the House' is taken from *My Way*, Leclair's first album recorded as Akufen. Containing over 2,000 samples, it was released in 2002 by Force Inc. Music Works, the Berlin-based imprint of Mille Plateau – named after Gilles Deleuze and Felix Guattari's *A Thousand Plateaus*.

7 In a review of albums by both artists, Michaelangelo Matos drew attention to their similarities: 'Both Edwards and Akufen. . .make house music from dippled 'n' dappled microsamples, creating collages from dozens of sources per track: concatenated horn bursts, a quarter of an inhaled breath, half an mmmm, dewdrop keys, clicky stuff, glorious syllable-splashes, instrumental Alka Seltzer fizz, hybrid micro-melodies, vowel needlepoint' (2003).

8 The roots of UK garage can be traced to 1992–1993 when US garage was being spun by DJs in the second rooms of clubs at jungle nights. As a darker side of the jungle scene alienated upwardly mobile young ravers, and females in particular, promoters and DJs began to focus on garage music for those looking to avoid drug-related violence. For more on this history, see Reynolds 1999.

9 In the US, Edwards has a lower profile and, according to one journalist, was 'recognised as just another producer in the enormous house pantheon, paling in hype to the prolific Todd Terry and the extremely in-your-face Armand Van Helden' (Host 2002, p. 19). A contributor to Daft Punk's album *Discovery* (2001), his profile has grown since the release of their album *Random Access Memories* (2013) on which he performed and co-wrote the song, 'Fragments of Time'.

10 Nick Prior explains how sampling technologies are used to disrupt, dislocate, deconstruct, and de-contextualise the voice: 'Filtered, chopped, stuttered, looped, repeated, mashed, reversed, pitched-up, pitched-down, degraded, resampled, sliced, quantised, warped, garbled, glitched, bit-reduced, timestretched, synced, mapped and tracked. These are just some of the actions and states that vocal samples undergo as a result of their transcription into binary code' (2018, p. 134).

11 In US law, *de minimis non curat lex* ('the law cares not for trifles') is an exemption that enables copyright material to be used without permission if the use is thought to be insignificant or the original work is unrecognisable in the newly created one. It has been used to argue that a small part of a recording or composition can be sampled without copyright being infringed. US judges have interpreted it differently. See Théberge 2004 for a discussion of this issue in the *Bridgeport v. Dimension Films* (2002) case and Latham 2003 in the *Newton v Diamond* (2002) case.

12 The two-part structure of 'Saved My Life' reflects twin tropes within US house: 'the metal machine music of the 'track' and the gospel humanism of the 'song'' (Eshun 2000a, p. 78).

13 Lyrics refer to love but Edwards describes the movement from track to song as a metaphor for spiritual awakening: 'That track is about a man who's going on a spiritual journey and he finds God. In the beginning in the music it's very chaotic sounding and then all of a sudden there's this crash and a gospelly sounding choir comes in. That's the point where he found God' (quoted in Host 2002, p. 21).

14 Kodwo Eshun describes the original as: 'a sombre, down-home blues sample with a vibrant hook of gospel chorale. Todd Edwards' remix was UK underground garage before it had a name, extracting vowel sounds that were stretched enough to register but so transient that they teased and

tugged, then crosshatching them with curlicues of guitar that licked your ear' (2000a, p. 80).

15 DJ and producer Matt 'Jam' Lamont has explained the role of Edwards' music in the development of UK garage: 'The most popular producer (American – it was almost all American then) was Todd Edwards. He put more *skip* [my italics] into his drums, changed the vocals round, and cut them up. . .When British producers started making their own music, they'd take the drums and the cut-up vocals, and push the bassline up a bit' (quoted in Benson 2000, p. 58).

16 Kenny 'Dope' Gonzalez is most well-known as one half of Masters at Work (MAW) with 'Little' Louie Vega. Along with Todd Terry, Gonzalez was one of the few producers working in US house and garage music during the 1990s to merge elements from house and hip-hop music.

17 The Ensoniq Performance Sampler (EPS) was introduced in 1988 and cost £1,695 in the UK. At the highest sample rate (52.1 kHz), the sample time was 4.95 seconds. See Anderton 1988a.

18 Schloss describes quantisation as the process that 'automatically moves samples to the nearest appropriate beat within a scheme that the producer chooses. For instance, if the producer chooses a framework of straight sixteenth notes in a particular tempo, the quantise function will set the beginning of every sample to the nearest sixteenth note. While this has the benefit of precision, it could, in fact, make the sequence overly precise or mechanical sounding' (2014, p. 140).

19 Edwards has spoken about retaining the 'imperfections' of 'live' instruments: 'If you truly want to make your stuff sound like it's not sequenced – like some computer just did it – it doesn't hurt to go in and do it by ear. Don't just let the computer quantise it, move it yourself. Sometimes it sounds good when something's off. It makes it sound like you're playing a live instrument, which I'm all for. I use all technology in my style but I still like it to sound imperfect. Imperfection is what makes it come across as pleasing to the ear' (quoted in Host 2002, p. 20).

20 Akai's S5000 and S6000 samplers were upgraded versions of its S-series samplers with a larger monitor-like interface. First announced by Akai in 1998, the S5000 cost £1,799 and the S6000 £2,799. Both contained 8MB of RAM memory, which could be expanded to 256MB (White 1999).

FIVE

Appropriation, Additive Approaches, and Accidents

Found

Introduction

In this chapter, I use material from an interview with a group of Edinburgh musicians and visual artists called Found who combine the writing of pop songs with the sampling of found sounds. I wanted to find out how Found use sampling devices like the Akai MPC2000 and whether they continue an art school tradition of making pop music by experimenting with new technologies.[1] My aim was to explore how and why Found's musical priorities have moved away from the appropriation of pre-existing recordings towards a recontextualisation of found sounds that is influenced by their study of art. Expanding on themes explored in the previous chapter, I was keen to discover if the core song-writing partnership of Ziggy Campbell and Tommy Perman used sampling instruments to sculpt sound in a similar way to how they paint or express their ideas in visual art. As well as examining *what* sampling technologies they use and *why* they use them, I wanted to find out *how* artists who are also musicians use sampling technologies. Subsidiary research questions were developed before I interviewed Campbell and Perman to gather data about the contexts of use and the process of using found sounds: What does the artist's studio look like – if it is a single place – and how important is sampling to the song-writing process? What is prepared beforehand in terms of melody, lyrics, and song structure, or is everything made in the studio? To what extent are sampling technologies used as compositional tools that form part of what Brian Eno (1983) calls 'an additive approach to recording' (p.57)? Before answering these questions, I begin with some information about the band, focus on what sampling technologies they use, and outline some of the influences that have shaped their approach to how they use them.

Finding Influences, Defining Sampling

Found began making music together in 2001 or 2002 depending on whether you read the biography on the band's website or their own record label's

119

website. A few years later a journalist described how their music 'blends bubbling dancefloor introspection with textured folk pop' (Robertson 2006). Formed by friends, Campbell, Perman, and Kevin Sim, while studying at Gray's School of Art in Aberdeen, the band have undergone a number of changes in personnel and been involved in a variety of art/music projects. These include their first two full-length albums, *Found Can Move* (2006), which they released on their own label, Surface Pressure Records, and *This Mess We Keep Reshaping* (2007), which was released by Scottish independent label, Fence Records. Catalogue numbers also extend to launch parties, documentary films, exhibitions, and performances. One event called 'Flight Path' (2006) involved members of the audience throwing paper airplanes through a laser beam to trigger sounds. Other projects include Cybraphon (2009), a custom-built musical instrument and 'emotional robot band', which expressed its feelings in songs determined by the amount of chatter about it on social media sites. What makes this case study different from the others in the second half of this book is that rather than working primarily as individual producers or artists like Marc Leclair, Todd Edwards, or Matthew Herbert, Found approach the use of sampling technologies as part of a more collaborative process of music making.[2]

As individuals in a small group, Campbell and Perman have approached digital sampling from different perspectives and this became apparent in answers about the sampling technologies they use and their admiration for particular users/musicians. For Perman, a love of hip-hop introduced him to sample-based music and he spoke about his early attempts to use cassette decks and four-track recorders to 're-create that sampling aesthetic of looping stuff up' (Perman 2008). He moved on to using Cubase[3] software before pooling resources with his brother, Bobby, to buy an Akai MPC2000[4]:

> I bought an MPC2000 with my younger brother. We went halfers on one. I learned my way around it then he basically took ownership of it somehow. I think he bought me out. I also had a [Boss] Dr. Sample [SP-202] for a while but I've used things like Cubase, which has a really good software sampler that I totally got into, and then Ableton Live, which is really an advanced sampler itself.[5] Everything that you do in that I consider sampling and manipulation. There is a software sampler in it called Simpler and a drum machine sampler called Impulse. The entire programme is just one big sampler as far as I'm concerned because of the way you can trigger loops and sounds and all the manipulating capabilities. They're all derived from things like the MPC and the E-mu SP-12.
>
> (*ibid.*)

Perman maps a relationship of continuity between hardware devices like those designed by E-mu in the 1980s and the development of software samplers like Simpler. When I asked him why he chose an MPC2000, the answer partly related to contingency:

> Just because somebody was selling it and needed to make some money quickly, and me and Bobby got it for a good price. I had used one before and at that point Kev [Sim] was using the MPC2000. Although it's a bit of a behemoth in terms of size and weight it's a really rugged, solid machine and it's just really user friendly. The thing that the Akai samplers are known for are the drum pads and that kind of instant touch is really conducive to programming nice drum beats and drum patterns. They've got a few features, which, once you start playing around with them, you realise that's how some of your favourite producers were doing things. You're just like, 'That note repeat thing – no way – that's so good'. And just pitching the 16 pads on the MPC2000, just pitching the notes, one sound over the 16 pads, you don't play it like you would play a keyboard. It opens up new ways of writing a melody or a bassline or something that you wouldn't do on a guitar or keyboard just because of the way it's laid out, so I find that quite interesting.
>
> (*ibid.*)

Campbell added: 'It's really playable. The velocity sensitive pads are unparalleled. It just really feels like you're playing an instrument' (Campbell 2008). The user draws a distinction here between a sampling technology – in this case, a non-keyboard-based hardware device like the MPC2000 – and a musical instrument. Yet, his physical relationship with the MPC2000 transforms it from a technology that is interpreted as something to be *used* into a musical instrument that can be played *with*.[6]

As well as Found's music making activities being shaped by hardware sampling/sequencing technologies like the MPC2000, the laptop is also important to their practices. Perman explained why he enjoys using software samplers to make music:

> We use a lot of software sampling now in Ableton Live. They'll have filters on them and as soon as you start playing around with the frequency filter you'll bring out sounds you didn't realise were in the recording you just made. You start playing around with the attack and decay of where your little looped sample is and suddenly, within seconds,

> you've created an entirely new sound you didn't think of
> before. I get really excited about that. I love the process
> and doing that with things I've sampled off a record or
> stolen online. You originally hear a horn sample or some-
> thing but, by the time you've put it in and played it on the
> keyboard, it's become something completely different.
>
> (Perman 2008)

As with Marc Leclair and Todd Edwards' microsampling practices, the
identity of samples is often unimportant and may end up as unrecognisable. Found also experience drawbacks, however, when using software
on a laptop or computer to organise samples into new recordings because
of 'affordances' relating to the quantisation of sounds:

> On the computer you're using a combination of your
> eyes and ears, which is interesting and it's faster for some
> things. It also means that you make music slightly differ-
> ently. On the MPC it's probably more intuitive. It's more
> about what feels right in terms of looping something up,
> like patterns that might have little bits that are slightly
> out of time, which just works better. On the computer
> there is a tendency to lock things into a quantised grid.
>
> (*ibid.*)

In the same way Edwards programmes his drum sounds so they sound
'less robotic', Perman wants his music to 'swing' like the music of hip-
hop producers Jay Dilla or Madlib. He discovered that this could also be
achieved using software samplers:

> A lot of the best programs these days have swing
> settings or swing quantisation. Live's got a master
> swing level that you can set up and it is so interesting
> listening to the difference that moving a drum hit a micro-
> millisecond makes to the character of the rhythm. It can
> suddenly make something sound infinitely cooler than it
> did when it had that kind of military effect.
>
> (*ibid.*)

Perman uses software samplers to create the swing that Akai's MPC
range is known for.[7] This user avoids the rigidity associated with
quantisation by editing and re-arranging sounds at the microrhythmic
level. While 'perfect' timing is made possible through quantisation, he
uses sampling/sequencing technologies like the MPC2000 and software samplers to replicate the sounds of humans playing acoustic
instruments 'imperfectly'.

As Perman's use of technologies shifted from making looped recordings on cassette tapes to using the MPC2000 and editing sounds on Ableton Live, his approach to the use of these technologies was shaped by both trip-hop and hip-hop producers and those whose sampling practices are not based around the re-use of pre-existing recordings:

> In terms of sampling it would be people like Geoff Barrow and the Portishead sound, which I discovered at the same time as a lot of hip-hop. DJ Premier and everything that he did I loved. What DJ Shadow was doing on *Entroducing* (1996) and the records leading up to that were hugely influential on me in terms of listening to music. It opened my ears up to a lot of pretty weird prog rock. Then, latterly, people like Matthew Herbert who I find very interesting. He writes very highly structured melodic songs and always has a very detailed approach to his production method. He's also got his own manifesto. It's very conceptual. And then Prefuse 73 was a huge influence with his *One Word Extinguisher* (2003) album, which I listened to to death. It is so melodic and built entirely on an MPC. That was a real eye opener as to what that technology could do.
>
> (*ibid.*)

Herbert has spoken of his frustration with approaches to sampling based around appropriation and may not value the listening skills involved in identifying a small segment of music to be looped and manipulated.[8] Perman, however, thinks that, in the case of DJ Premier and Pete Rock, 'Their ear is like a good photographer for finding that loop, which is five minutes into a rare jazz tune. Suddenly, there's a lick that was in an improvised solo, never repeated, and for whatever reason they've sampled it' (*ibid.*). Its dismissal as a lazy form of plagiarism can ignore the complexities of an approach to the use of sampling technologies that began to frustrate Perman. He describes how he

> . . .fell out of love with the idea of stealing other people's stuff, partly because it's so difficult. You have to really work a sample to turn it in to your idea or manipulate a drum pattern, which can be great fun. Then you start working with a drummer. We sample our own drummer and chop up the samples. I get him to play in the studio. Then I go away and process that drum component a hell of a lot to achieve something half way between a live recording and a homage to my favourite hip-hop or dance music producers.
>
> (*ibid.*)

Perman's experience of these difficulties involved in the recontextualisation of sampled sounds led him to a process of recording 'real' musicians and producing a hybrid of 'live' and 'recorded' music. Along with Herbert's ideas around the non-use of pre-existing recordings, this is the reason why he moved away from sampling pre-existing sounds and began looking for other sound sources to shape his musical practices.

While Perman was moving away from an appropriation-based aesthetic, his fellow band member, Campbell, was moving closer towards one. When first introduced to digital sampling by the third member of Found, Sim, his initial experiments involved sampling his own guitar playing. He then started to sample sounds from pre-existing recordings and began buying second-hand records specifically for this purpose:

> I'm not really from a hip-hop background so the first I knew of sampling was when I met Kev and he was telling me about this box he had, which was just for DJs. It was a really simple sampler, a Vestax one. It didn't do that much, you could pitch with it. The first thing I started thinking of was how you could sample yourself. I wasn't that interested in sampling other people. I've actually become much more interested in that.
>
> (Campbell 2008)

His music making was also influenced by the music and ideas of artists like John Cage and Steve Reich, which he was exposed to while studying at art college: 'I like the concepts and I like reading about these guys. Sometimes more than I like the music' (*ibid.*). While Reich's tape loops are often cited as a precursor to the looping of recordings using sampling technologies, it is Cage's ideas relating to the non-distinction between music and everyday sounds that are more relevant to Found's musical practices. They incorporate 'the sounds of everyday life' into their music but, unlike avant-garde artists, want to retain the rhythms, melodies, and harmonies of pop music. Campbell and Perman converged on the position that sampling technologies did not have to be used solely for the appropriation of pre-existing sounds. They began using them to record, manipulate, and loop sounds from any sound source.

One of the ways in which Campbell and Perman use sampling technologies is in line with the early design objectives of technologists at Fairlight Instruments and E-mu. Where the users of the Fairlight CMI sample libraries were limited to a generic range of musical instruments, the users of software packages like Apple's Logic Pro can access the pre-recorded sounds of specific keyboard instruments from a much larger library:

> I have within my laptop now, thanks to Logic, samples
> of hundreds of instruments, particularly keyboards and
> drum kits, which are actually samples, not synthesized
> versions of them. So, there's instrument sampling where
> you re-create one of the first uses of the sampler. It's not
> the same as playing a Fender Rhodes Mark II Suitcase
> Piano or whatever but I can't afford one and I don't
> know anyone that's got one. It gives me the opportu-
> nity to sketch something out with a Fender Rhodes and,
> often through a little bit of filtering and a couple of
> effects here and there, it sounds great. As a sketching
> tool, it's absolutely fantastic and that's sampling pos-
> sibilities at a very low cost. It opens up a massive range
> of musical instruments that you can use. I don't swear
> on it. I don't particularly like the final piece being played
> that way. I've got opinions against it but in terms of
> composition it's fantastic and so that's another area of
> contemporary technology that's really helped me.
>
> (Perman 2008)

In the early 1980s, users of digital synthesizers/sampling technologies gave
mixed reviews about how successfully they could replicate the sounds of
acoustic or electric instruments. As a user of software samplers, Perman has
no issues with their fidelity levels. However, he is uneasy about using a laptop to imitate the sounds of certain electric instruments and there remains
a sense that it would be better to use the 'real' thing for live performance.

As well as using sample libraries, the members of Found sample the
sounds of acoustic and electric instruments to create a library of sounds
that can be played on the keyboard. Perman explained this when defining
sampling as a process involving any digitally recorded sound, not just the
appropriation of sounds from pre-existing recordings:

> From my point of view, the definition of sampling is the
> process of recording a sound, having it in what would be
> a sampler, a bit of technology, that can play back a sound
> on demand. It doesn't matter where that sound source
> comes from. You can then manipulate the sound. We
> always manipulate the sample no matter if it's off a record
> or if it's a guitar strum. That, or the sort of ring after you
> play a guitar chord, becomes a really interesting keyboard
> sound once put into a sampler and played over the octaves.
> I think that's definitely sampling. Or you would take your
> voice, pitch it, and manipulate it. That's sampling.
>
> (*ibid.*)

The process of manipulation is key to an ideology of creativity around sampling that places emphasis on what happens *after* the act of digital recording: it is not just about taking sounds but transforming them to create new sounds or new ways of using sounds. In the first chapter I explained how composers at the BBC Radiophonic Workshop used the Fairlight CMI to create new musical instruments by juxtaposing digitally recorded sounds: Clarjang was made from a clarinet sound and a metallic noise. Perman uses his computer, software, and keyboard instrument to 'mix' sounds together: he samples a snare from a pre-existing recording and combines it with the crackle of vinyl. He mixes the sounds of a clarinet with the sounds of falling rain after recording outdoors:

> We were down at Edinburgh Sculpture Workshop and recorded sounds there. For me, one of the nicest moments was when we had this guy who was playing the clarinet in this little pavilion with the rain beating down on it. He held some really long clean notes and, as soon as he'd gone, I fed it into the computer and started playing chords with them. It just sounded so nice. There was something so woody and organic about it with the rain crackle in the background.
>
> (*ibid.*)

Rather than using digital editing tools to remove them, the group incorporate the sounds of 'the natural world' into recordings. They record the playing of acoustic instruments like clarinets in makeshift studios and then use computing technologies to create new sounds that are celebrated as 'organic' even though they have been processed digitally.

Found Sounds: Appropriation, an Additive Approach, and Accidents

I now want to look at three recordings by Found to illustrate three different approaches to sampling that they use in their music: appropriation, an additive approach, and the inclusion of sounds derived from accidents or unplanned events in the recording studio. One of the few Found songs with a sample from a pre-existing recording can be heard on 'Some Fracas of a Sissy', a track from *This Mess We Keep Reshaping*. They use a short sample of a trumpet from the song 'Night Life in Shanghai (Ye Shanghai)' by Chinese singer Zhou Xuan throughout the first half of the track. Campbell explained:

> It's a straight lift. I liked it because it wasn't like we were taking a groove from it and building a whole tune. I knew

> we'd taken horns and a little bit of the female vocal and
> then when I built the track I thought, 'Fuck it. I'm just
> going to let people hear where I've taken this from'.
>
> (Campbell 2008)

After the second verse, the short trumpet sample is followed by a longer
sample of Xuan's vocals. According to Perman, revealing the source of
a sample in this way is a common tactic among hip-hop producers. In this
case, the members of Found assumed the source was an obscure Chinese
song and did not clear the sample.[9] As a result of its use during the Beijing
Olympics in 2008, they realised it was actually a well-known song in China.[10]
Despite this, the copyright owners have taken no legal action over their use
of the song and recording, which is partly explained by Found's semi-
professional status, the release of their recording on a small micro-label,
and China's historically lax copyright laws.[11]

Found began to view the appropriation of pre-existing recordings as
marginal to the sampling practices of the group – the reasons given for
this are artistic rather than legal – and a more typical example of their use
of sampling technologies can be heard on 'Static 68', a track from their
first album, *Found Can Move*. The starting point for this recording was
not a melody, lyrics, or musical ideas lifted from a pre-existing recording.
Their focus was on using sounds normally considered extraneous to the
processes of listening to vinyl records, though Perman wanted to avoid
being taken too seriously:

> I don't want this to sound too pretentious, but it starts
> with an atmosphere or a character rather than chord
> structures. There'll be a sample that has got a bit of dirt
> to it that's really interesting. 'Static 68' started out as a
> record static loop, which I then built lots of stuff on top
> of. It was the static from the run-out groove that was the
> most interesting thing for me and the whole song grew
> around that.
>
> (Perman 2008)

The original sampled sound becomes irrelevant and, even listening very
carefully, it is difficult to hear the sound of static in 'Static 68'. This
approach to the use of sampling technologies maps directly on to what
Brian Eno describes as an additive approach to recording where sound
production technologies can be used 'to chop and change, to paint a bit
out, add a piece' (1983, p. 57). Found's sampling practices are continu-
ous with ideas about the recording studio as a compositional tool and
earlier uses of technologies to 'paint with sound'. Perman makes the comparison with 'an abstract painter in the way that they keep on working

a canvas until they're content and sometimes the original under painting is completely lost' (Perman 2008). Whether it is the sound of static or a truck reversing, they may remove these digitally recorded sounds at the end of the process, despite being, what Perman calls, the 'initial seed' (*ibid.*) of the song's idea.

As well as using the sounds of rainfall, static, and truck engines in recordings, Found also employed sampling technologies to retain and experiment with the sounds of urban life that have been recorded unintentionally. Campbell experienced problems caused by the noises of everyday life in tenement buildings when trying to record in a home studio:

> Something I'm getting into just now through necessity because I've got a really noisy neighbour is just not waiting. I used to wait for quiet times to record, especially vocals because you have to have the mic quite high, but now I'm just going to leave it all in. If there's someone cutting the grass or a dog barking, I'm just going to leave it in there. It's the same aesthetic as the snare drum thing [discussed earlier]. It's a kind of sampling, unfettered sampling.
>
> (Campbell 2008)

While digital editing tools are often used to remove unwanted noises or mistakes, the solution here is to record, sample, and experiment with them to add an authenticity to recordings. An example of this in Found's music can be heard on 'See Ferg's in London' from *This Mess We Keep Reshaping*. Campbell's recording of the final vocal take was interrupted by a phone call that caused him to leave the room. Rather than edit out accidental noises in the attempt to capture a 'perfect' performance, he left in the sound of the door opening. This is different to the short interlude in Kate Bush's 'All We Ever Look For' where the opening of a door leads the listener to the sounds of a fictional performance. For Perman, it provides what he calls a piece of 'punctuation' and a short pause in the music that occurs just after the lyric 'when everything's gone quiet'. The interruption to the recording also acts as a reminder to the listener: this is a 'live' performance; this is *real*.

Sampling and the Home Studio: Dislocated Recording/ Live Performance

In *Any Sound You Can Imagine* (1997), Paul Théberge describes the growth of home studios in the 1970s as a private space for performers to try out musical ideas before entering professional studios to record them. There is still a tendency to talk about 'going into the studio' and I imagined Found's melodies, lyrics, and song structures being constructed as part of a two-stage process of writing and recording. However, Campbell explained that everything occurs in the one place: 'It's all done in the studio really. When we say in the studio we mean in our bedrooms but it's not like we prepare demos and then go and do it proper. It's all part of the same thing' (*ibid.*). Each musician in the group defines the studio differently depending on the equipment they use. For Campbell, the home studio is a misconception; depending on domestic arrangements, it can comprise of a much smaller area of private space. Albin Zak refers to examples of 'location recording' (2001, p. 105) when rock bands like The Band or The Red Hot Chili Peppers left large state-of-the-art recording studios in favour of old mansions in Los Angeles with primitive mixing desks and outdoor swimming pools. The way in which Found work might be described as an example of *dislocated recording* where each member of the band usually works alone in domestic or temporary spaces before coming together at different points in the process when their physical presence is required. Their use of laptops equipped with recording software and software samplers enables collaborative practices to occur flexibly without having to go into a place designated as a professional recording studio.[12]

While Campbell's contributions are performed and recorded in his bedroom, Perman explained that he and other band members can be even more flexible in terms of the spaces they use to create music: 'Kev's studio is his sampler and he'll just sit with his headphones plugged in so he can work anywhere. My studio's just built around my laptop with a few things plugged into it' (Perman 2008). The studio Perman uses as an artist and the one he uses as a musician involve the same hardware technologies: 'I sit at my laptop and kill my eyes for music and art. I'll sit and draw at the very same desk where I'll write crap little melodies. I'll switch between having to do something in Photoshop to working in Ableton' (*ibid.*). The boundaries between the role of musician and artist are as fluid as the flick of a cursor or the prerequisites of funding applications. When I asked if Found consider themselves musicians or artists or both, Campbell answered that 'it depends who's paying. It depends what we're applying for' (Campbell 2008). The definition of a recording studio is just as fluid, consisting solely of a single piece of hardware (a laptop) with software (Ableton Live) that enables sounds to be digitally recorded, stored, and manipulated.

When I interviewed Found in 2008, they were keen to move from the dislocated recording experience described above to one where they could capture the experience of their live performances by 'playing together' in a professional recording studio.[13] Campbell thinks 'that something special happens when people play music live together' (*ibid.*). He values the isolated use of sampling and other sound recording technologies – individual acts of music making with non-human technologies – less than the 'shared oxygen' of two or more humans making music together as part of a live performance.[14] For Found, though, this does not exclude the use of laptops and sampling technologies, which are as important to their live performances as drums, bass, guitar, and keyboard. They noticed the absence of a dedicated sampling/sequencing device during one performance when Sim, who programmes and plays the MPC2000 on stage, did not turn up on time because he had overslept. Their set list options were limited, managing to perform five songs without him before 'he appeared in the crowd and everyone cheered. It just made everything make total sense and have more impact' (Perman 2008). Sim had previously used the MPC2000 as a click track for the band to keep time when performing live but, when used with a delay pedal, it became an instrument of live improvisation:

> The two of them together means Kev can have so much variation in his sound. He'll trigger a sample then mess with the settings on the delay pedal and he can really change the pitch. It's a different live performance every time he uses it and that, for me, is an example of a really good live musician.
>
> (*ibid.*)

While artists like Stevie Wonder used digital synthesizer/sampling technologies such as the Fairlight CMI as part of live performances in the early 1980s after their transportation by aeroplane or car, samplers and sampling technologies have often been thought of as studio tools that are used (rather than played) by a single individual. The smaller size, design, and mobility of sampling/sequencing technologies such as the MPC2000, or laptops with software samplers, means they are not just compositional tools used in recording studios. They are musical instruments used as part of live performance, collaboration, and real-time improvisation. Through both their design and use, they have also contributed to re-shaping the definition of what a recording studio is and can be. In the case of Found, the laptop might be described as a 'boundary object' (Star and Griesemer 1989) containing software for cultural production in the fields of art and music that moves between and enables users to work in different social worlds.[15] While this and other technologies like the MPC2000 are used with enthusiasm, what remains is an older discourse of authenticity around ideas of playing live and using *real* instruments.

Notes

1 For more on experimental practices in art school education, their influence on pop musicians in Britain since the 1960s, and the use of technologies by art-school trained musicians like Brian Eno, see Frith and Horne 1987: '[I]n the 1960s art school students became rock and roll musicians and in doing so inflected pop music with bohemian dreams and Romantic fancies' (p. 73).

2 'Individual' artists collaborate too, of course. As Howard Becker writes: 'All artistic work, like all human activity, involves the joint activity of a number, often a large number, of people' (1982, p. 1).

3 Cubase is a music software package developed by Steinberg to record and sequence audio and MIDI data. In 1996, Cubase 3.02 was released with the Virtual Studio Technology (VST) interface and plug-ins. Thom Holmes writes: 'The VST specification encouraged the widespread development of plug-in instruments, effects processors, and MIDI controllers' (2016, p. 509).

4 Akai launched its range of sampling drum machines with MIDI sequencing in 1988 with the MPC60 MIDI Production Center. Designed by Roger Linn – his company Linn Electronics had closed in 1986 – it was modeled on the Linn 9000. Akai released an updated version, the MPC60 II in 1991, which was followed by the MPC3000 in 1994 and the MPC2000 in 1997.

5 In 1999, Gerhard Behles and Bernd Roggendorf started a company called Ableton in Berlin. Robert Henke, who produced electronic music with Behles as Monolake, had been developing hardware controllers for live performance and they began working together on the music software package, Live, released in 2001. For an interview with Henke on Live, see Kirn 2011.

6 Mark Katz describes the transformation by hip-hop DJs of an object – the turntable – into a musical instrument as 'a process, and this process requires not a single individual, but an entire community' (2012, p. 62). Objects have to be *socially accepted* as musical instruments.

7 Akai's MPC series is associated with the use of swing settings to programme non-quantised rhythms in hip-hop and electronic music. Roger Linn attributes these 'natural, human-feeling grooves' to a number of factors including the drum pads and note repeat function on the MPC range. The swing settings were first developed on the LM-1 Drum Computer and referred to as 'shuffle'. For an interview with Linn on the topic of microtiming and MPCs, see Scarth 2013.

8 In an interview in 2006, he stated: 'With a sampler there's no distinction between sound and music, or noise and music, and I think that's a liberation that musicians have struggled to find for years. We finally have it and instead people are using it to rip off their record collections, which confuses the hell out of me' (quoted in O'Neil 2006).

9 Perman told me how his experience of clearing samples left him determined to never go through the process again: 'If we sample something then we'll try and hide it or we'll just not worry about it. What would happen is we'd get a cease and desist letter and that'd be it, like Danger Mouse['s *The Grey Album*]. That was a high-profile case where it made his career and there was absolutely no downside to that project whatsoever so I was an idiot to try and clear stuff' (Perman 2008).

10 For more on Zhou Xuan and Chinese popular music, see Stock 1995. Stock writes about how the Chinese authorities rehabilitated the music of 1930s Shanghai and the recordings of singer-actress Xuan in the 1980s and 1990s as a response to the circulation of US and European pop music on cassette tape. In 1985, China Record Company reissued *Jin Sangzi Zhou Xuan* (*The Golden Voice of Zhou Xuan*) followed by the collection, *Zhou Xuan*, in 1993.

11 For more on micro-labels and the practices and discourses of their owners, see Strachan 2007. For more on recent changes to Chinese copyright law and government policies relating to the impact of digitalisation on its creative industries, see Street, Zhang, Simuniak, and Wang 2015.

12 Nick Prior explains how mobile music technologies have extended 'the possibilities of collaboration and iteration. For instance, band members no longer have to be physically co-present to collaborate with each other. Software files and audio files can be easily sent through electronic or regular mail to be added to, modified, or mixed, then returned for further iteration' (2008, pp. 919–920).

13 This occurred with their third album, *factorycraft* (2011), which was recorded in Chem19 Recording Studios and released by Chemikal Underground.
14 On musicians and non-human technologies when overdubbing in recording studios, Albin Zak writes: 'Overdubbing requires the performer to summon up inspired performances in the absence of not only an audience but other musicians. What in a live situation is an interactive interchange among players – a kind of musical breathing together – becomes a one-way responsive relationship between the musician and a fixed, unchanging musical partner, the track' (2001, p. 54).
15 Like 'interpretative flexibility', the concept of a 'boundary object' is derived from the social study of science. Star and Griesemer describe them as 'objects which both inhabit several intersecting social worlds *and* [original emphasis] satisfy the informational requirements of each of them (1989, p. 393).

SIX

Foot Pedals and Folk Music

King Creosote

Introduction

In this chapter, I explore how a device not usually identified with sampling has been used to re-shape practices and sounds within a genre traditionally opposed to the use of electric technologies. With many of its musicians and ideologues continuing to value the authenticity of 'unmediated' performances, folk becomes an interesting site of study for understanding contemporary approaches to the use of digital technologies.[1] While this ideology of authenticity became less important as it splintered into sub-genres, musicians who are associated with folk and use digital technologies still have an ambivalent relationship with the genre.[2] This has been evident among members of the Fence Collective in Fife, Scotland, including Kenny Anderson (aka King Creosote) whose music has been labelled both 'new folk' and 'indie folk'.[3] Anderson released albums on its micro-label, Fence Records, with songs including found sounds, lyrics from email conversations, and extracts from voice-mail messages.[4] Choral voices from classical recordings can also be detected in the low-fidelity mix along with instruments more tradition-ally associated with folk (acoustic guitars, accordions). To investigate these incongruities, I met with Anderson and a semi-structured interview was guided by three basic research questions: What digital sampling technologies do you use? How do you use these technologies? Why do you use them? The answers to these questions provide the structure to a case study about the use of a foot pedal as a sampling device. It focuses on a user of sampling technologies who is ambivalent about their use as well as his relationship with folk.

Old Instruments, New Folk: Accordions and Foot Pedals

Anderson's association with folk music is partly due to family connections – his father is the accordion player and ceilidh band leader Billy Anderson who has presented a radio show called *Sounds Scottish* on Tay AM since the 1980s. It also relates to his membership of the folk/bluegrass band, the Skuobhie Dubh Orchestra in the late 1980s and early 1990s.[5] Rather than considering himself part of the Scottish folk scene, however, Anderson views his life as a performing musician with the Skuobhie Dubh Orchestra and his development as a singer/songwriter (King Creosote) quite separately:

> As somebody that played folk music on an accordion when I started in bands, it was almost like starting a different musical life entirely. The two things were no way connected. When the Skuobhie Dubhs started up, that was a bluegrass, kind of folk band. I wasn't writing songs for them so writing my own music and recording it was completely different.
>
> (Anderson 2009)

Anderson's musical identities may be more fluid than this account implies: the accordion is one of the instruments he uses when performing as King Creosote. For this reason, he is often labelled as a folk musician, which sits uncomfortably with him:

> It rankles me a bit when people hear the accordion and instantly classify it as folk when to my ears it's obvious that the songs I write are not folk songs. They're just normal pop songs but the fact that I've used an accordion or sing in a Scottish accent has somehow made them folk songs. The folk songs I know and love tackle age-old worldly themes. They do it well, and that's what makes them as relevant today as when they were written in the 1800s or whenever. I don't see my obscure little songs being relevant in twenty years never mind in 200 years.
>
> (*ibid.*)

Though he wants to write pop music, Anderson has enjoyed little chart success and has been described by one journalist as an 'alt-folk darling to the (relative) masses' (Wilson 2009). He wants to distance himself from folk music, but this is not because his use of technologies has proved contentious with the ideologues who police the genre.

Anderson's early songwriting experiments as King Creosote and his sampling practices were shaped by a single device: a second-hand guitar pedal.[6] He bought it in the early 1990s when the Skhuobie Dubh Orchestra went in search of new musical equipment:

We never bought effects. We never bought amps. And
then for some reason we decided we should get some
of these musical toys. We went over to Dundee and
it was like, 'Maybe it would be good if the fiddle had
an amp. Maybe it would be good if I used a couple of
guitar pedals'. I was just looking at guitar pedals and I
bought an equaliser. All it does is give you a bit of EQ
and then I saw this second-hand pedal. It was a digital
delay pedal. The guy in the shop did the salesman pitch
and said, 'Look it's a really good quality pedal and it's
pretty cheap'. So, I bought this pedal but, because we'd
gotten so used to not taking any additional equipment,
all this stuff just ended up sitting about.

(Anderson 2009)

Another reason for their non-use was that some members of the band
had stricter ethical ideas about the kinds of instruments that were appropriate for folk and bluegrass music. According to Anderson, they also
faced aesthetic challenges when trying to combine the sounds of acoustic
instruments with those produced using a digital delay device:

With the Skuobhie Dubh Orchestra the hardest thing was
getting the other band members to let these things in. The
drummer was very forward-thinking and he was up for
using anything but at the other end of the scale I had my
brother [Een] who would just be like, 'How can I play a
banjo over that?' or 'Banjos are for this style and this style
only'. He has always been a bit of a purist when it comes
to the sounds that you make, the instruments that you use,
and the style that you play. I'm a musical tart really. I just
put anything together. My main drive was to find things
that sound new and combinations that sound new.

(*ibid.*)

Instead of being shaped by folk ideologies, Anderson's ideas about mixing
sounds are influenced by a concept with origins in the art music world:
originality. He uses the delay pedal as a sampler to create music as part of a
modernist aesthetic; one that is focused on the development of new sounds
rather than the preservation of traditional songs.

Anderson bought his delay pedal with the intention of using it to create
effects for his guitar but, after some initial experiments, decided not to
use it in this way. The pedal became a different kind of musical object:
'I didn't use it all. It was just something I bought as a digital delay pedal

and then it was too much hassle to set it up, to set a delay time. I never used it as a pedal. It just went in the box and, in fact, I've never used it as a guitar pedal. It's my sampler' (*ibid.*). Anderson's discovery that the pedal could be used as a sampler to digitally record, store, and loop sounds was accidental. It was the result of 'mucking around' with things and asking questions about how they work:

> It's a dead simple pedal – two pots to twist and a switch to push with your foot. One pot is marked with different time settings, the other you could set to sample, once or repeat. For years I never used it with my acoustic guitar. Then one day I was mucking around with it and discovered that a hit-or-miss combination of sample and repeat would give me up to four seconds of a loop, which I could pitch using the time pot. I was like, 'Wow!'.
>
> (*ibid.*)

Anderson did not mention the make or model of the pedal during our interview. In later email correspondence, he explained that the foot pedal was appealing because of its colour – 'the same metallic green as dodgem cars' (Anderson 2013) – rather than any design specifications. In a comment that can be interpreted as light-hearted rather than assigning agency to a non-human object, he said, 'It chose me' (Anderson 2009). While sampling technologies in the 1980s were used in ways that were unforeseen by their designers, this is a pedal that was used in ways that were unanticipated when buying the device.[7]

Working in a home studio without other musicians or much equipment, Anderson rediscovered the pedal when he was searching for sounds to add to recordings:

> When The Skuobhie Dubh Orchestra broke up and I started recording as King Creosote, I used a digital eight track, a Casio keyboard, my guitar, accordion, and some percussive bits and bobs. I was always looking out for inexpensive gadgets to help modify my small pool of sounds. That's when I found the second-hand digital delay pedal, no manual or box of course, and was like, 'Oh aye, what's this?'.
>
> (*ibid.*)

As mentioned in an earlier chapter, J.J. Jeczalik ignored the Fairlight CMI manual and preferred to learn about the instrument through a process of trial and error. Even if the delay pedal had come with a manual, Anderson would have likely ignored it as well:

I've got a real phobia of manuals and I've got a real
phobia of having to sit down with a bit of kit for hours
to work it out. Every time you use it you have to plan
ahead. It's maybe just laziness but I really find a lot of
music done in that way so boring, like using Pro Tools
and all that cutting, pasting, and looping. To me, that's
more of a mathematical, computer-based exercise.

(ibid.)

Despite rejecting folk as a label, Anderson's ideas about the creative
process are shaped by its ideology – he has an aversion to new, in this
case digital, technologies. As a songwriter, he wants the process to be
artistic rather than overly technical and the song to be separate from
its digital editing. This, though, assumes that the artistic expression
of ideas can be disconnected or separated from the technologies used
to express them.[8]

   With a fear of using technological devices as part of 'a technical exercise' (ibid.) or thought processes deemed to be too rational, it seemed
unlikely Anderson would want to use a rack-based sampler as part of
his music making. However, I asked him if he had ever been interested in
buying a device from Akai's range of digital samplers:

No, it's just too much bother. The drummer in my band,
Gavin [Brown], uses samplers and he's continually asking me to find the sources of samples I've used so that
he can programme them into his Akai. Even when we've
done that, to my ear, they don't sound right because
they are right. He's made them right. He's given them a
start point and an end point. He's almost gotten them
bang on in time with the song. He's either time-stretched
them or he's made them how they should have been but,
to my ear, that sounds wrong. What I like about the
samples I've used and the fact I'd gotten used to them
is the irregularities in them. All the wrongness that is
in those samples has become the thing that I hear and
when I hear it sanitised, cleaned up, or quantised, it just
doesn't sound right.

(ibid.)

Simon Frith has written about how musicians are more likely to use the
language of right and wrong rather than good and bad when considering musical decisions (2004, p. 19); this is the case with Anderson. He
wants his recordings to contain the kind of 'mistakes' and 'errors' that
can be removed using digital editing tools. While Found and other Fence

Collective members have used Akai's MPC range of digital sampling/
sequencers, Anderson has never had the urge to buy, play, or experiment
with one:

> Not really. I've watched other people working them and
> I'm just like, 'I don't know how you can be bothered
> with that'. It's too clinical. It's too 'music by numbers'.
> 'Oh, you've got to set this. You've got to set that. You've
> got to do a crossfade. You've got to take the start and
> the end and get rid of any clutter'. No, you don't.
>
> (Anderson 2009)

With fixed ideas about music making as a particular kind of creative act,
Anderson thinks that Akai's devices are instruments to be programmed
rather than played. Devices dedicated to digital sampling do not appeal
to him and he favours using music technologies to create 'imperfections'
as part of the DIY ethic of 'new folk' music.

Sample Sources and the Non-Use/Use of Delay Pedals in Live Performance

Anderson's discovery that his delay pedal could be used as a sampler led
to a search for sounds to record and loop. He began trawling through an
old collection of cassette tapes:

> I was like, 'This thing acts as sampler'. I started going
> though old tapes that I had, listening for sections of
> music that were uncluttered. I wanted things that were
> stripped back so I was going through classical-esque
> tapes and old random stuff that people had given me.
> I'd known about sampling and that The Verve were
> sued for sampling The Rolling Stones [on 'Bitter Sweet
> Symphony'] so I knew there was some illegality around
> it but, at the same time, I reckoned it'd be okay if I went
> for more obscure and classical stuff that's out of copy-
> right. So, I was going through folky things. I was going
> through classical things, opera things. . .
>
> (ibid.)

There are similarities here between Anderson's sampling of pre-existing
recordings and the technological practices of loop-based hip-hop. Yet it is
a genre he has little interest in: 'It's something I admire but it's just a style
I'm not into. I've never heard a hip-hop record and thought 'wow'. It's just
something I've never listened to. I don't have any hip-hop records' (ibid.).

While there was an unwritten rule among US hip-hop producers in the late 1990s about sampling only from original, rare, and often expensive, vinyl recordings (Schloss 2014), Anderson does not adhere to any strict rules or a set of sampling ethics. He samples sounds from a more random selection of sources: old classical tapes, music gifted from friends, and CDs given away free with music magazines such as *Q*. The sources of Anderson's samples are whatever happens to be 'lying around'.

Though Anderson is not interested in searching for breakbeats on vinyl records, one of his first attempts to use the delay pedal as a sampler was to create a loop from a group whose music was sampled heavily by hip-hop producers in the late 1980s and 1990s:

> The first proper loop I sampled was of drums and electric guitar from a song intro on an old untitled cassette mixtape. I just set my pedal to four seconds maximum and caught the first beat of the bar – it just so happened that the four second maximum length was exactly a bar and a half of what turned out to be a Funkadelic classic. . .like, bang on, couldn't be better. I was working on a song called 'So Forlorn' at the time and I was able to play every chord in my song on top of the sample without a single note clash. I had a four-chord trick with a slightly off-kilter three note guitar riff and drums behind it, in time, and in tune. In this way, I had Funkadelic as the backbone for the whole song, returning to the original source to nip out little 'woo ooh' vocal whelps and so on.
>
> (Anderson 2009)

On the version of 'So Forlorn' from the album *Kenny and Beth's Musakal Boat Rides* (2003), the looped sounds from the Funkadelic recording are multitracked with Anderson's acoustic guitar and accordion. The source of the sample is not as important as whether the sounds complement the chords in the pre-existing song. Without any instructions on how to loop sounds using the delay pedal, he attributes serendipity to the process of synchronising sounds from the pre-existing recording with the recorded sounds of acoustic instruments. Rather than possessing specific programming skills that he has taken time to learn or read about in a manual, Anderson suggests it is about luck.

On another song from *Kenny and Beth's Musakal Boat Rides* called 'Turps', listeners can hear a sample from a pre-existing recording of classical music. Anderson's decision to use this was not just because it was (assumed to be) out of copyright. There were also practical reasons including the difficulties of recording instruments in a home studio:

> I was soon using sampled choral works instead of layering up my own backing vocals, or strings pitched low in place of a poorly recorded bass guitar. I've used a lot of unusual world music sounds from magazine-mounted CDs. I wasn't necessarily looking for drum loops, guitar wig outs, or anything like that. It was just to capture an accidental performance from a random source. I soon had an ear for stripped back music sections that I could pitch and twist into use, and for a few years I used that sampling pedal as a versatile yet chaotic instrument. I got into sampling my own vocals, previous recordings of my old bands, anything that came to hand.
>
> (*ibid.*)

While the sources of Anderson's samples tend to be from a random range of pre-existing recordings, he has also appropriated performances from unreleased studio recordings by fellow members of the Fence Collective. A series of drum sounds and solos recorded in a coal cellar in the early 1990s doubles as an ad-hoc sample library and a source of drum loops. On the sleeve notes of the King Creosote album *Red on Green* (2004), Rich Amino is thanked 'for any drum loops that fell off his recording on to mine'. Rather than music lawyers being employed to negotiate the clearance of rights to use ideas and sounds, any possible issues over copyright infringement are dealt with less formally or not at all.

Anderson mainly uses his delay pedal while working on recording sessions in his home studio and tends not to use it while performing live. This is because of the difficulties of using sampling technologies in real-time to incorporate pre-recorded sounds into songs, especially when a paying audience is watching and listening to a structured performance:

> It's hard to play live with samples because the samples I've used are not clear. They're quite muddy sounding. You can't really hear what's going on. When those looping pedals came along I did have the notion of getting one of those because it's a similar way of doing things. You've got to punch in and get out. It's very live but then I was thinking, 'Where would I get my samples from? Would an audience sit there and wait while I fast-forwarded and rewound a tape to get a bit, nip it out, pitch it, no that's not quite right, go back?' I didn't see how that would work in a live setting.
>
> (*ibid.*)

The other reason for Anderson's non-use of the foot pedal on stage is because he thinks listening to a recording and going to a live performance should be two different experiences: 'I'm not into replicating recordings at gigs anyway. I think, 'Great, leave all those songs sample-ridden because that's how they were done' but the live version doesn't have to have the same thing' (*ibid.*). The use of looping pedals like Anderson's have become common in the live performance of 'new folk' and other genres of popular music by artists like Ed Sheeran and KT Tunstall. Tunstall famously used Akai's E2 Head Rush pedal when performing on the TV show *Later. . .with Jools Holland* in 2004 and creates a groove during solo performances by tapping a rhythm on the side of her guitar with her hand and recording, then looping, the sound.[9] Anderson dismissed this approach to using looping pedals as a technological gimmick when I asked him about his own non-use of the delay pedal during live performances:

> I find it quite dull when people use those pedals and overuse those pedals. They all do the same sort of thing. They tap their guitar. They play a little lick on the guitar and you're limiting yourself to chords that'll work over those notes. With most of the people I've heard using those pedals, the song ends up the same. You get this over-saturated mush and they all do it. It's a shame but that put me off really. I don't know if I want to go down that same route because everyone's doing it the same kind of way. I don't see the point of that really.
>
> (*ibid.*)

Preferring to sample the sounds of pre-existing recordings, Anderson is keen to position himself as an artist who uses technologies in ways that are unpredictable, unplanned, and not part of a social trend. His aesthetic is about difference, which fits neatly with the 'indie folk' ethos of the Fence Collective and an ideology of not 'following the crowd'.

Colour, Texture, and Off-kilter Randomness

Anderson inserts samples into his music at a point in the recording process where something is needed to provide more colour or texture. His approach to multitrack recording chimes with the 'additive approach' discussed in the previous chapter:

> In most cases I've done the guide guitar, done my percussion, and put a bit of keyboard on. I've only got eight tracks so I'm getting to about track six, or seven,

or eight now, and thinking, 'I just need something else:
what is it?' That's the point where I listen to things and
see if could nip out either a choral thing, or strings, or
anything. It really depends at what point I need to have
that extra thing.

(*ibid.*)

Anderson's use of sampled sounds is a way of adding more instruments
or sounds to the mix to create something 'a little bit different'. Even
though he did not study at art college, his discourse is shaped by the same
tradition as artist/musicians like Found:

For me, adding samples is adding colour. I'd say 90%
of my samples just add a texture that I cannot get from
the few instruments I have lying around the house. It's
so much easier to nip out three notes from a bar of
Beethoven and repeat them ad infinitum than to cobble
together a string section, dot the music, [then] get them
rehearsed, and recorded well.

(*ibid.*)

Anderson uses the delay pedal to 'paint with sound' and add colour to
recordings but with less sophisticated tools than those used by Found.
For him, the delay pedal is an energy-saving device. He uses it to loop the
sounds of instruments on pre-existing recordings rather than co-ordinating
a process where 'real' musicians can be recorded playing them.

In contrast with the aims of instrument designers at Fairlight
Instruments or New England Digital, Anderson does not want to have
more control over sound. The use of the foot pedal is important because
it introduces a randomness and 'off-kilter' feel to his music:

With the delay pedal you've got no control over how
long your sample is. You can't just cut in and cut out.
You've got to try and guess how long you want your
sampling window to be but it's dead easy to pitch it once
you've got it. What's not easy is to get it to continually
loop. I used to say that my samples were either in time
or in tune but never really both. What that did with my
music was it threw up a kind of random element. I quite
liked the idea that these samples were all off-kilter with
the song and your ear had to grow to kind of learn the
patterns. It adds a little bit of playability to your songs.

(*ibid.*)

A deliberate decision to leave things unpolished and create a tension between sounds considered 'right' and 'wrong' also extends to the volume of particular samples:

> Some samples just did not work if they were too loud. If I got them quiet enough they seemed to add a kind of weirdness, even though they were probably in the wrong key. In fact, I know a lot of them were in the wrong key but if you had it quiet enough certain notes would just disappear. I haven't yet heard anyone else use samples in quite the same way because there's something really uncontrived about it. It just sounded really naïve. Now, when I listen back, it just sounds like two musical styles coming together by chance and working in a random way. I still hear new things and new combinations of notes when you've got this weird, off-kilter sample running up against something metronomic. Every time it comes in it's completely different.
>
> (*ibid.*)

Anderson's desire to create 'a unique bit of music' (*ibid.*) separates him from the folk ideology of preserving tradition, which he is ambivalent about. However, in describing his use of the delay pedal as 'uncontrived', he highlights how the discourse of authenticity, which was based around the use of 'real' sounds and 'real' instruments in the first two case studies, continues to be important to him and the users of sampling technologies. The importance of authenticity in folk has disguised how it has always been mediated by technologies: the notebooks and pencils of Cecil Sharp, the portable tape recorders of John and Alan Lomax, the multitrack recordings of folk rock artists in the 1960s and 1970s, and the sampling devices used by Kenny Anderson and other members of 'new folk' music scenes in the first decades of the twenty-first century.

Notes

1 See Frith 1981 for more on the ideology of folk as a celebration of 'pre-capitalist modes of music production' (p. 159). As archivists, collectors, and folklorists used sound recording technologies to document folk traditions in the twentieth century, this ideology became more problematic.
2 Rob Young writes that, when used in reference to contemporary music, folk has become 'as much a signifier of texture and aesthetics as an indicator of ingrained authenticity – as in such descriptive terms as 'acid folk', 'free folk', even the ungainly 'folktronica" (2010, p. 8).

3 Described by one journalist as 'the Fife indie-folk mafia' (Maxwell 2009), Anderson started the Fence Collective in the 1990s with artists like KT Tunstall, James Yorkston, and The Pictish Trail. For more on its history, see Galloway 2013. For a history of 'new folk', see Encarnacao 2013.

4 Anderson has released more than forty albums as King Creosote. Two albums were released on the major-owned label 679 (*KC Rules OK* (2006) and *Bombshell* (2007)). Albums released by Domino Records include *Kenny and Beth's Musakal Boat Rides* (2003), *Flick the Vs* (2009), and *Diamond Mine* (2011) with Jon Hopkins, which was nominated for the Mercury Music Prize.

5 The Skuobhie Dubh Orchestra is pronounced Scooby Doo Orchestra after the cartoon character. The band, whose members included KT Tunstall and Anderson's brother Een, released three albums but broke up when their relationship with a record label based in Scotland 'started to go a bit sour' (quoted in Cloonan, Frith, and Williamson 2003, p. 110).

6 Effects units have traditionally enabled musicians to alter the amplified sounds of various instruments. Foot pedals (also known as stompboxes) have been used to create a range of sonic effects for guitarists. Steve Waksman explains that pedals are 'small metal boxes containing transistor circuits that, when connected between the line that ran from guitar to amplifier, altered the electronic signal delivered to the amp, changing the sound. The most common such device was the distortion-inducing fuzzbox, a staple of [Jimi] Hendrix's sound' (1999, p. 183).

7 In the 1980s, companies like Boss and Ibanez designed pedals that could digitally record, store, and loop sounds. For a useful history of guitar effects pedals and more on the introduction of digital delay pedals, see Hunter 2013: 'The wonders of digital delay arrived on the pedalboard in the early 1980s with what seemed massive capabilities for long delays, clean signal reproductions, and the endless fun of one, two, or up to 16 seconds of looping delay' (p. 38).

8 As Frith writes, 'The 'industrialisation of music' cannot be understood as something which happens *to* [original emphasis] music, since it describes a process in which music itself is made – a process, that is, which fuses (and confuses) capital, technical, and musical arguments' (1992, p. 54).

9 John Richardson explains how the pedal enables real-time simulation of multitrack recording. He writes: 'Contrary to the received wisdom on looping and other repetitive practices, which portrays them as 'passive' and when, sampling is involved, 'parasitical', the evidence of this television footage [of Tunstall's 2004 performance] points towards a heightening of agency through the performer's immersion in the act of composition' (2009, p. 91).

SEVEN

The Sounds of Everyday Life (and Death)

Matthew Herbert

Introduction

My final chapter of Part II focuses on the music and ideas of Matthew Herbert. His practices are important to my argument about diverse approaches to the use of sampling technologies because he samples from everything *but* pre-existing sound recordings. A set of rules he produced called the Personal Contract for the Composition of Music (Incorporating the Manifesto of Mistakes) prevents him from doing so. He avoids all pre-existing sounds including those on synthesizers and drum machines. There is also a rule about using 'real' instruments rather than virtual instruments. Like early users of synthesizer/sampling instruments such as the Fairlight CMI, Herbert wants to use 'real sounds' (Herbert 2012) and, for this reason, he uses sampling technologies to make field recordings. In this chapter, I explore Herbert's use of found sounds in dance music – food being digested, knuckles being cracked, teeth being brushed – and his use of field recordings made in sewers, war zones, and crematoriums. Using data from an interview with Herbert, I situate him in the field as a user of digital technologies who places restrictions on his sample sources to develop a more 'authentic' approach to sampling.

House Music + Musique Concrète = House Musique Concrète?

The musical practices around field recordings remain associated with art and folk worlds rather than the worlds of popular music.[1] As a contemporary field recordist, Herbert has a complicated relationship with the musical and technological practices in genres like pop and hip-hop. He is also ambivalent about the dance music scenes that he was connected with in the early stages of his career. Though he speaks disparagingly of contemporary pop music and 'the sort of trancey r&b bullshit that's coming out of the charts at the moment' (*ibid.*), he has composed music for the Eurovision Song Contest. While admitting to being influenced by hip-hop producers like DJ Premier and albums such as De La Soul's

Three Feet High and Rising (1989), Herbert was never interested in exploring the loop-based approach to sampling that has dominated the genre. He denigrates the appropriation of pre-existing sounds as a form of musical consumption: 'I wouldn't have described it then in these terms but in retrospect I would say that it feels like consumerism. It feels like shopping, musical shopping. It's like 'music by numbers' and you just happen to choose the numbers. It didn't seem that interesting to me' (*ibid.*). Instead, Herbert prefers to collaborate with orchestras like the London Sinfonietta and remix Gustav Mahler's *Symphony No. 10* for Deutsche Grammophon by inserting the sounds of glitch and electronic dance music (EDM). His art music credentials were underlined in May 2012 when the BBC (in partnership with The Arts Council of England) re-established the BBC Radiophonic Workshop as The New Radiophonic Workshop (NRW) and he was appointed its Creative Director. Herbert positions himself in the field as an idiosyncratic user of sampling technologies who moves between the worlds of art, classical, avant-garde, *and* popular music without being central to any of them.

Herbert's interest in manipulating sounds began at an early age and, like Tommy Perman of Found, it started with the use of analogue technology as a cut-and-paste tool. He recorded sounds from the radio in a family home that contained tape-based technologies, which his father had a professional interest in and expertise about how to use them:

> I don't know whether I inherited it from my Dad, who was a sound engineer at the BBC, but there was a certain interest and love of technology in the house. I had tape recorders at home that I'd always record bits of the radio and the Top 40 with. I'd chop up little bits and take bits out that I didn't like. That's from the age of about nine or ten. The music shop was a very exciting place to go as a twelve or thirteen-year-old with digital samplers being invented and the prices coming down. I don't think it was a surprise, or a coincidence rather, that I started to get more into it as soon as I was able to buy something.
>
> (*ibid.*)

The first sampler Herbert bought was a Casio FZ-1 in 1995 – 'It was one megabyte. It was the size of a table' (*ibid.*) – and cost him approximately £300. Despite having up to 29 seconds of sample time available, living in rural Kent placed limitations on the sources of sound he could sample from. He was not able to sample sounds from pre-existing sound recordings because there were no shops nearby where he could buy records and he made no mention of raiding his parents' record collection. Instead, Herbert began to make music from the sounds of more mundane objects:

> The first thing I sampled was an apple. Then it was a pepper pot. Then it was just things lying around the house so books, video cassettes, and radios. It was a very domestic version of music, quite literally. It was whatever was to hand. What I sample now is completely different as is the principle behind it and my way of working. At the beginning it was: I'm not quite understanding the potential of this tool and sampling anything.
>
> (*ibid.*)

In my interview with him, Herbert admitted to having used sounds from sample library CDs but said, 'I'm totally against that now. I'm totally embarrassed about it' (*ibid.*). For him, sound libraries now represent a technological quick fix he disassociates himself from; he now prefers to sample the sounds of everyday life, or what he calls 'real sounds'.[2]

Herbert's move towards the use of field recordings as part of his musical practices can be traced back to his use of found sounds in the production of dance music. In 1998, he released *Around the House*, an album that revealed his interest in the sounds of domestic spaces. It begins with the sounds of crockery being cleared up; an intercom buzzer signals the prelude to a romantic night-in: 'Hey honey, come up'. The special guest is co-writer, and now ex-wife, Dani Siciliano who sings gently in the background.[3] The connection between the songs on the album and specific rooms in the house is not always clear from the floor plans or the titles, many of which refer to difficulties in the personal relationship of the couple ('We Go Wrong', 'Never Give Up'). The sounds of sex or snoring are absent from 'Bedroom Jazz'. The most significant scene of this domestic drama takes place over the sounds of breakfast 'In the Kitchen'. In an interview, Herbert stated:

> At the beginning you hear me saying, 'Right, what ingredients have we got? Beans, bacon, eggs'. From start to finish we make breakfast. What you hear is us laying out the plate. There's one point where you hear me cutting open the bacon, that's really loud. The toast is going *tiktikertikertiker*. We start frying things. That's when you hear *shsshhhhhurrur*. Then we get round to eating it and there's a loud noise. That's me spreading butter on toast. I picked that one sound, used that as one of the percussion sounds, picked out a couple of other nice

> sounds and layered them through. From start to finish
> the whole process is in real time. There's no edits. From
> walking into the kitchen to taking it out and eating it –
> that's how long it takes, which is not very long, about
> 11 minutes.
>
> (quoted in Eshun 2000b, p. 37)

A plaintive song called 'The Last Beat' suggests this is a tour through a
relationship with insoluble problems. 'Going Round' soundtracks love
as non-linear but there is little to indicate these emotional experiences
are taking place in specific spaces. This is an album of house music
with Herbert using 'real' instruments such as guitar, bass, and piano
and introducing ideas from *musique concrète*. Some of his later concept albums are more coherent and composed entirely from the digitally
recorded sounds of 'everyday life'.

Herbert's fascination with found sounds and field recordings was developed further on the album, *Bodily Functions* (2001). This time around
his palette was created by sampling the sounds of the human body including 'the blood of Martin Schmidt' (Herbert 2013a) on 'Foreign Bodies'
and the sounds of laser eye surgery on 'You Saw It All'. Schmidt is one
half of the electronic duo Matmos who are credited as engineers on these
two tracks. They released an album in the same year called *A Chance
to Cut is a Chance to Cure*, a concept album about cosmetic surgery.[4]
Herbert's project was less macabre and along with the sounds of food
being digested, knuckles being cracked, and teeth being brushed, there
are samples from new-born babies, the slamming of doors, and the recycling of bottles. Journalist Rob Young wrote that the album 'harked back
to the very earliest post-war tape experiments of *musique concrète*: Pierre
Henry's creaky tape cut-up *Variations For a Door and A Sigh* is the clear
antecedent for much of his early music' (2003, p. 26). The comparisons
with art music composers neglects how much of Herbert's earliest music
was designed for dance floors. He has, though, been keen to avoid being
typecast as a dance music producer. A biography written for his website
explains how he

> . . .would later distance himself from this early work, in
> that he felt a little too deeply implicated in the hedon-
> istic club scene of the time but primarily because he
> had sampled other people's music, for which he would
> later be repentant. 'I feel it is a betrayal of what I really
> believed to be the right thing to do at that time. I was
> seduced and shaped in part by people and assumptions
> around me'.
>
> (Stubbs 2012)

His denunciation of dance music and club culture, or dance music associated with mass culture, reads as an attempt to position himself as a musician whose practices have little relationship with popular culture. This can also be attributed to the manifesto he drew up in 2000, the Personal Contract for the Composition of Music [P.C.C.O.M.], which shaped the making of *Bodily Functions*. It has also placed creative constraints on his use of sampling technologies and the music-making projects he has been involved in since.

Contractual Obligations: 'The Sampling of Other People's Music is Strictly Forbidden'

Herbert created the P.C.C.O.M. after being inspired by the Danish filmmaker Lars Von Trier and his Dogme collective. They launched a manifesto in 1995 announcing their decision to use only hand-held cameras and avoid props, special effects, or the addition of sound during the post-production process.[5] As with Dogme's manifesto, Herbert's contract for the composition of his own music consists of ten self-imposed commandments along with an optional rule relating to the remixing of recordings by other artists. It prevents 'the use of sounds that exist already' (Herbert 2005a), including the use of drum machines or factory pre-sets and pre-programmed patches.[6] The contract also states that '[n]o replication of traditional acoustic instruments is allowed where the financial and physical possibility of using the real ones exists' (*ibid.*). Above all, the sampling of other people's music is not permitted. Herbert explained that the introduction of a personal contract was a reaction to the way in which newly released (digital) technologies were being used at this time and a desire for the music making process to be more difficult:

> There was a strong movement in technology to encourage you to start writing music in a certain way, to always take the easy route, and to take the short cut. Actually, the most exciting thing was that music could now be documentary. You can make and take real sounds as opposed to something that already exists and the personal contract was to remind myself to do that.
>
> (Herbert 2012)

The ideology of authenticity that runs through this book, in the discourse of designers and users of sampling technologies, reappears in Herbert's rules. In his case, there is an enthusiasm for the creative possibilities available to musicians through the use of digital technologies to record and manipulate the sounds of everyday life. At the same time, there is a need to preserve 'real' sounds and 'real' instruments in his music.

Like Dogme's approach to filmmaking, Herbert's contract was an attempt to challenge what he viewed as conventional music making practices that began to develop with the use of digital technologies, particularly around the removal of mistakes. His contract states: 'The inclusion, development, propagation, existence, replication, acknowledgement, rights, patterns, and beauty of what are commonly known as accidents is encouraged' (Herbert 2005a). He explained why he was keen to avoid perfectionism:

> One of the important things about accidents is that they undermine the more traditional elitist perspective of the auteur, the composer, the genius, the maestro, and all the male ideals of hierarchy. For me, there's a political aspect to it, which is handing control to something else. The second thing is it feels much more like a human experience. It feels like you're part of the process rather than opposing the process.
>
> (Herbert 2012)

His argument against auteurs hides Herbert's desire to achieve the status of a virtuoso in the field of sampling. When I asked if he played any other instruments, he replied:

> I do but they all have limited interest for me because if I pick up the guitar I can't play it as well as Marc Ribot or Jimi Hendrix. If I play the piano I can't play it as well as Thelonious Monk or Bill Evans. But if I've recorded the sounds of London sewers, no one else has got that. There's no one else to compete with.
>
> (*ibid.*)

There was humour in his response and Herbert does not take himself too seriously. However, his individualism is in contrast to the way a collective like Found operates with a number of core members and different musical roles. He credits himself as both the composer and copyright owner of sounds from 'the natural world' that have been digitally recorded. These are projects conceived by Matthew Herbert on which many people provide support. This includes other users of digital technologies who have helped him in the field/s to capture sounds that have been too difficult for one person to record.[7]

Herbert makes his field recordings with portable tools including Sennheiser MKH 418-S microphones and a Nagra V 24-bit Linear Location Recorder.[8] For recording and manipulating sounds inside the studio rather than out in the field, Herbert admitted to buying and using almost every sampling technology that becomes available:

I'd buy one of everything and try and use them. The only ones I've never owned are the Akai S950s. I've never really liked the sound of them actually. The one that I use a lot is the Akai S612, which is the first one that Akai ever made and just does one sound. That's what I use live. A huge amount of my live sound manipulation, my whole live career, has been built around one of those because it's the only sampler that has a start and end point, mover or slider, so you can instantly cut a sample up manually. It's also got filters and knobs. There's nothing internal so it's much more like an analogue synth and much more playable than some of the others we've had to programme. So, I've used virtually everything and ended up using the Casio FZ-1 for years. I used it for fifteen years and then in 1999 I got an E-mu, the e64,[9] which was pretty great. I'd had an E-mu SP-1200 drum machine for a while, which was pretty great. I had an Akai MPC, which I didn't get on with.

(*ibid.*)

Interestingly, Herbert highlights the playability of Akai's first rack-based sampler but did not develop a successful relationship with the MPC range, which the members of Found considered to have an intuitive interface. He also expressed frustration with samplers that are designed for users to build a library of sounds rather than search for 'new' ones:

After a couple of years, I abandoned sampling other people's music and started to sample my own sounds. There's very, very few samplers that were set up to do that. For example, they were never designed to fill them up every time you used them. They were always built thinking you'd use the same sounds again and again. You'd create a library and you'd use that. Akai's samplers have always been bad apart from the S612. They've always been really bad at quick manipulation and quick input of a lot of samples. The MPCs and the S950s took hours to truncate one sample. It would take you ten minutes to get one sample ready. If you're sampling fifty times a day every day for ten years, you just want it to be as quick as possible. Now most of my recording or sampling is of sources outside of the studio, which I import into the computer. Now I just use soft samplers, which I don't really like but the capacity to store 100 gigabytes of recordings that I've made is really great.

(*ibid.*)

Herbert's use of software samplers does not conflict with any of his self-imposed rules so long as he does not use any pre-sets or software synthesizers. In the earlier chapter on microsampling, Marc Leclair expressed enthusiasm for being able to record for hours using software samplers. For Herbert, the benefits are also to do with the options they give him for labelling, organising, and archiving sounds in the memory of his PC.

With software samplers, Herbert indexes sounds using a keyboard, monitor, and mouse. For him, this is much quicker than the time it takes to scroll through the alphabet to label a track or a sound using a sampler with a small LCD[10] screen like an Akai S612:

> The good thing about software samplers is their ability to handle huge amounts of data and for it to be catalogued as well. It sounds like a really, really small point but being able to give sounds a name is really important. My library of sounds contains over a million sounds now. Those old LCD screens where you scroll through the entire alphabet, first in capitals, then in lower case, just to be able to change the first letter, slowed you down enormously. When you're trying to work quickly and trying to get a musical idea going, it's not really happening. One of the reasons for using software samplers is that the indexing is so much quicker.
>
> (*ibid.*)

Herbert has access to the manuals for the samplers he owns but does not read them. He employs others to perform tasks with software packages that he does not enjoy using:

> I've never had the patience to read manuals. I always think it's a failure of the technology if you have to use the manual a lot. Another really important thing is I had time then. When I first started I was unemployed but now I don't have the time to learn new technology. I once started with Max/MSP[11]. I bought it and I spent a few hours on it, didn't get enough back from it, and I've never used it since. Now I just pay someone to build the patches for me. There just isn't the time to learn new instruments and new technologies. It's exhausting.
>
> (*ibid.*)

Herbert is both a user *and* non-user of music technologies. He outsources tasks perceived to be less artistic and important to his technical assistants or, what Howard Becker calls, 'support personnel'.[12] Like Trevor Horn, who owned a Fairlight CMI but employed J.J. Jeczalik to learn how to use it, Herbert is a user of sampling technologies who wants to be recognised for his cultivated approach to their use but does not always have time to learn how to use them himself. He created a personal contract to make the process of using digital technologies more difficult at a time when they could be used to do things quicker. Yet as a user/non-user, he is not averse to designing his own solutions to save time.

The Sampling of Politics: Moral Rights and Wrongs

Since introducing his personal contract, Herbert has used digitally recorded samples as the basis for many of the sounds on his albums and, occasionally, *every* sound. All sampled sounds used on these albums are listed meticulously on his website. For his Radio Boy project, *The Mechanics of Destruction* (2001), he recorded the sounds made by squashing Big Mac meals and ripping Gap boxer shorts and re-composed them as music. On his jazz big band projects, *Goodbye Swingtime* (2003) and *There's Me and There's You* (2008), he sampled the sounds of items like supermarket tills, a guillotine, and a passport. Some albums are based around a single concept such as the Wishmountain album *Tesco* (2012), which was made solely from the sounds of the UK's top ten best-selling brands in 2010. A preoccupation with food, consumption, and politics also led to him to make *Plat du Jour* (2005), an album containing compositions such as 'The Truncated Life of a Modern Industrialised Chicken' and 'Nigella, George, Tony, and Me'. On the latter, Herbert recorded the sound of a tank reversing over the meal that Nigella Lawson served for George Bush and Tony Blair in Downing Street while they discussed the invasion of Iraq (Herbert 2013a). *One Pig* (2011) incurred the disapproval of PETA (People for the Ethical Treatment of Animals) for attempting to record the sounds of a pig's life at every stage; these included its difficult birth in a sty and its eventual death six months later in an abattoir that Herbert was not allowed to enter for legal reasons. It is not always clear *who* Herbert is recording these sounds for: how many listeners are likely to endure the unpleasant and disturbing sounds on these concept albums from beginning to end?[13]

Herbert wants to expose his audience to sounds they might not normally hear and gained access to a crematorium where he recorded sounds not usually heard by mourners:

> In the crematorium where they burn the body for about ninety minutes, possibly two, or two and a half hours, there's still bits of bone left. They scrape the bones out. They put them in what's called a Cremulator, which is like a washing machine with some graphite balls

in it. The bones go around in this drum and it's the most fucking hideous noise you'll ever hear in your life. When I was there I recorded the sound. The guy's like, 'Record this. This is a great noise'. So, I recorded it. I got it home and added it to this remix that I did of Mahler's *Symphony No. 10*. Knowing what it was and the fact it was this hideous sound made it really disturbing and it brought a huge emotional weight to the piece. It gave it a weight of authority and horror that wouldn't have come from many other things. But if that was my mother's bones and somebody had recorded her bones being ground down and put them on a record without asking her permission, I'd feel pretty pissed off about it. I hadn't asked their permission. I didn't know anything about this person whose bones I had recorded so I took it out. The end result is not quite as good but, morally, it's much more appropriate than having the bones of an anonymous stranger ground down, even though it's the most extraordinary noise you'll ever hear.

(Herbert 2012)

Questions are raised about whether certain sounds should *not* be sampled. Herbert has broached this subject by sampling the sounds of Palestinians being shot by Israeli soldiers on *There's Me and There's You*. *The End of Silence* (2013) is 'made entirely from one six second recording of a Gaddafi war plane dropping a bomb during the battle of Ra's Lānūf on 11th March 2011 made by war photographer Sebastian Meyer' (Herbert 2013b). His music is politically motivated and he wants to present listeners with the mediated realities of war. The contexts are complicated, though, and it is worth speculating whether a code of conduct is needed for sampling musicians in the same way that journalists, photographers, and editors operating in theatres of war are obliged to follow guidelines about what can and cannot be broadcast, printed, or published.[14]

Herbert is trying to change the world with his music but he is also trying to change the way in which people hear and perceive the world:

I think that's true. I think I'm probably going to fail on both counts [laughs] but yeah absolutely. I don't know why I would aim for anything less. I don't particularly want to write a piece of music that's going to make people feel like drinking a bit more beer in the bar tonight.

(Herbert 2012)

In the 1990s, anti-capitalist and environmental campaigners used slogans about reclaiming the streets as part of their protests. Herbert, though, has a much wider aim:

> It's about reclaiming the world. It's about saying these things can also be musical instruments. You can turn shit into music. You can turn waste into something permanent. One of the proudest things about the *One Pig* record, which is not what I set out to do, is that this pig would just literally be poo and in landfill. Here we are, three years later, and somebody in Scotland, somebody in Russia, somebody in Australia, and somebody in America knows about and has listened to the life of that pig. It's a way of remembering. It's a way of creating little monuments to these events instead of letting them just go past unforgotten.
>
> (*ibid.*)

As a meat-eater, Herbert wants to reveal the 'uncomfortable realities of eating meat' (Herbert 2011) and create a musical memorial to what he calls 'the dispossessed and the unheard' (*ibid.*). In valuing the meaning of every single sample and sound, he uses the digital sampler as the basis for political statements about capitalism, globalisation, and consumerism. At times, the messages are didactic – for example, the liner notes to *Plat Du Jour* instruct listeners to avoid using supermarkets (Herbert 2005b).

Since the 1980s, the discourse around the ethics of digital sampling has been focused on issues of borrowing and intellectual property (Porcello 1991). Herbert's practices raise a different set of questions about the moral rights of technology users. These questions relate to the moral rights of human and non-human animals where the sounds of their bodies have been digitally sampled during or after their life: Who owns the digital recordings of these sounds? Who should be credited as the owner of these sounds? What permissions are required to use them? Herbert decided against using the sounds recorded in the crematorium because he was unable to ask permission from the family of the deceased. However, this was a matter of conscience rather than of copyright, which is not something he has to contend with at the moment:

> Not yet. It will be in a hundred years' time. McDonalds will have copyrighted the sound of their burgers. Mazda will have copyrighted the sounds of their MX5. Edinburgh University will have copyrighted the sounds

> of their campus. That will all come. We've got a window now. That's why I'm trying to dash through it as fast as possible and collect up the scraps.
>
> (Herbert 2012)

There are issues, though, when Herbert takes some of the more unusual or controversial sounds he has sampled out of their original contexts. It is important to him to make it clear where these sounds are from, as their careful categorisation online and in liner notes suggest. These sounds, however, may go unnoticed, appear random, or remain unidentified unless listeners read about the sources. I asked Herbert if there was a danger that some political meanings might be lost or the ability to make a statement about the situation in Palestine lessened if the sound of a gunshot is divorced from the location of its recording. He directed me towards his experience of composing idents for the Eurovision Song Contest and his attempt to challenge the visual and sonic representation of Israel while its pop stars performed to a global audience:

> I did the sounds in between the films when Russia hosted it in 2009. I had to come up with one for Israel and they had all these images of happy skateboarders. I was like, 'I can't just pretend that Israel is one happy skateboarding family' so I inserted the sounds of gunshots and Palestinian homes being bulldozed. You can still hear it. I can't believe I got away with it. You can hear it all. It's pretty great to be subversive on that sort of scale but I think the important thing is that it's music first. It should draw people in.
>
> (*ibid.*)

This political act is judged by Herbert to have been a success even though the samples embedded in the piece were *not* recognisable or identifiable by the intended audience. The question here becomes who the performance is for and whether the political message the music is trying to communicate is understood or oblivious to its listeners.

In *Any Sound You Can Imagine* (1997), Paul Théberge recognises that the use of digital technologies has helped to change our definitions of what music can be.[15] The use of digital samplers has also changed our definitions of what a musical instrument is and can do: they can be used to record, store, edit, manipulate, and reproduce a greater range of human and non-human sounds in increasingly complex ways. Herbert's practices also raise questions about the role of the musician and ways that it continues to change with the use of digital music technologies and tools.

He argues that the digital recording of sound now enables him to assume the role of a documentarian rather than a musician:

> I definitely think sampling has a historical quality to it and a historical purpose to it. This is what the world sounds like and in a hundred years' time, if people are still listening to music or listening to it in this way, they'll be able to hear what the sewers in London sounded like in musical form. I absolutely think it's a living diary in that respect. It's about bearing witness.
>
> (Herbert 2012)

To say that Herbert is recording the sounds of the world, however, ignores the high-tech manipulation of those sounds: it is not the sounds of sewers we are hearing but the combined sounds of sewers, samplers, microphones, and manipulation. As digital technologies become increasingly entangled in the social practices of musicians, the ideology of transparency and realism that has existed since the early days of sound recording continues to be part of the discourse of contemporary users of music technologies. Herbert's discourse and practices demonstrate his desire to capture 'real sounds' and the sounds of the human and non-human world in unmediated ways.

Notes

1 For a series of interviews about the practices and use of technologies by field recordists including Francisco López, Hildegard Westerkamp, and Felicity Ford, see Lane and Carlyle 2013.
2 In *Any Sound You Can Imagine* (1997), Paul Théberge traces the growing importance of sound libraries for users of synthesizers and digital samplers in the 1980s and quotes from a magazine review of Yamaha's SY77 synthesizer in 1990. It states: 'Producers and keyboardists will find that sound libraries are a must because it takes so long to programme 'real' sounds' (p. 81).
3 Kodwo Eshun described the album as 'a tour through a domestic landscape, each vinyl side's label presenting a floor plan showing the movements of the duo, amplifying the emotional contours of interior space' (2000b, p. 37).
4 Jim Haynes describes how the album was made using 'a wealth of field recordings from nose jobs, cauterising muscle tissue, laser eye surgery, and liposuction. The album begins with the light shuffling house groove of 'Lipostudio (And So On. . .)', which introduces an odd textual duet between human fat gurgling through a tiny vacuum and a bleating clarinet' (2001, p. 28).
5 For more on Von Trier and the rules of Dogme '95, see Roman 2001 and Simons 2003.
6 See Goldmann 2015 for interviews with instrument designers Robert Henke and Mike Daliot, producer Michael Wagener, and artist Cory Arcangel on the use of pre-set sounds.

7 Chris Pickhaver is credited with making additional field recordings on a project called *One Pig* (2011). On the front cover of the album *There's Me and There's You* (2008), Herbert lists the musicians, engineers, photographers, journalists, and web designers who have been involved in the project. It takes the form of a petition in which the undersigned agree that 'music can still be a political force of note and not just the soundtrack to over-consumption'.

8 The MKH 418-S is a shotgun microphone that enables sounds to be recorded in Mid-Side (MS) stereo. The Nagra V is a digital recorder with removable hard disk and can record one hour of 24-bit audio at a sample rate of 48kHz per GB of disk space.

9 E-mu released the e64, a rack-based sampler, in 1995. With 64-voice polyphony and 2MB of RAM, which could be expanded to 64MB, it cost £2,650 (Wiffen 1995).

10 LCD stands for Liquid Crystal Display.

11 MAX/MSP is a visual programming language based on two earlier programs: Max (named after Max Mathews) and MSP (after Miller S. Puckette). Thom Holmes writes: 'Max/MSP can trigger audio-processing routines at the same time that it manages other aspects of a performance, such as the spatial distribution of sound to loudspeakers, the triggering of MIDI devices, and the multitrack recording of the outcome. The time needed to master an audio development environment such as Max/MSP can be daunting' (2016, p. 326).

12 Becker writes: 'Participants in an art world regard some of the activities necessary to the production of that form of art as 'artistic', requiring the special gift or sensibility of an artist. The remaining activities seem to them a matter of craft, business acumen, or some other ability less rare, less characteristic of art, less necessary to the success of the work, and less worthy of respect. They define the people who perform these special activities as artists, and everyone else as (to borrow a military term) support personnel' (1974, p. 768).

13 In the liner notes to *One Pig*, Herbert thanks 'anyone who has the time and inclination to sit down and listen to the whole record from start to finish, in order, in one go. . .' (Herbert 2011).

14 The Society for Ethnomusicology has a Position Statement on Ethical Considerations (1998), which contains guidelines about carrying out fieldwork in a responsible way. On the use of recordings, it states: 'Ethnomusicologists acknowledge that field research may create or contribute to the basic conditions for future unanticipated, possibly exploitative, uses of recordings and other documentation. They recognise responsibility for their part in these processes and seek ways to prevent and/or address misuse of such materials when appropriate'.

15 He writes: 'Recent innovations in musical technology thus pose two kinds of problems for musicians: On the one hand, they alter the structure of musical practice and concepts of what music is and can be; and, on the other, they place musicians and musical practice in a new relationship with consumer practices and with consumer society as a whole' (1997, p. 3).

Conclusions

The starting point for my research about digital sampling was that its study had been dominated by a focus on the legal framework of copyright within the music industries. Concentrating on the re-use of pre-existing recordings in genres such as hip-hop had been at the expense of examining how samples have been used in other genres of popular music. In this book, I looked at a range of practices relating to sampling and the use of sampling technologies in a variety of musical worlds: pop, rock, hip-hop, folk, and electronic dance music (EDM). I demonstrated that sampling and its uses have often been defined too narrowly. Samplers have often been viewed as social weapons that were used by artists like Negativland and The KLF to attack the concepts of copyright and authenticity. By using empirical evidence, I explored how sampling technologies have also been used as musical instruments, editing tools, compositional tools, and social mediators. Music technologies are ways of mediating the world and some users are trying to reproduce its sounds in unmediated ways. As well as my emphasis on the contemporary uses of sampling, the first half of the book focused on the history of sampling technologies to understand how technologies, musical practices, and discourses have changed since the late 1970s.

To understand technical objects, Madeleine Akrich recommends studying the relationship between designers and users as well as the relationship between 'projected' users and 'real' users. This was the aim of my research and my findings show how sampling technologies have been employed by 'real' or actual users in ways unimagined by their designers. The 'real' or actual users shaped the re-design of digital synthesizer/sampling instruments as part of a feedback loop. Peter Vogel told me that users of the Fairlight CMI contributed sounds to its sample library. Fairlight Instruments published a newsletter for its users and also organised a User's Club. Further research on the amateur and semi-professional users of the Fairlight CMI would be helpful for understanding how their practices relate to those of the professional users who shaped the first half of the book. After following the instruments, designers and users in the first three chapters, I followed the users of music technologies in my case studies and highlighted the multiplicity of sampling technologies being used in the processes of contemporary music making. I showed how the 'interpretative flexibility' of sampling instruments was assumed to have closed in the mid-to-late 1980s and 1990s: a consensus was reached about their use as they became associated with the appropriation of pre-existing recordings. As was demonstrated by the case studies in the second half of the book, however, there has been no stabilisation or closure mechanism; there is still 'interpretative flexibility' as digital sampling technologies continue to be used for different purposes in a range of musical worlds.

The case studies in this book focused on how users within different social groups have been using sampling technologies over the last two decades. Rather than make generalisations about the use of digital technologies, I emphasised the diversity of musical and technological practices using a sample of semi-professional and professional users. The first case study focused on two EDM producers, Marc Leclair and Todd Edwards, who both explained the advantages of using software samplers and the amount of memory available for storing sounds. In the case of Edwards, it was the human voice that was central to the sounds he sampled and re-composed to create virtual choirs and spiritual messages. In the case of Leclair, he recycled microsamples from radio broadcasts and immortalised them as part of recordings he thinks of as collages. The second case study focused on Found, a group of semi-professional users of sampling technologies. For them, an Akai MPC2000 acted as a digital sampler, a sequencer, and a studio. Like the laptop, it is a meta-device.[1] Theirs was a case study about the problems of using digital sampling technologies in bedrooms and other types of home studios. It was also about the use of mobile music technologies and the way they enabled a *dislocated recording* experience that occurred in multiple 'studio' spaces rather than in a single location.

In contrast to the chapter about Found, which was based around a group of users and how they negotiated the use of hardware devices and software samplers to make music, the third case study returned to an individual user – Kenny Anderson (aka King Creosote) – who works as part of a loose community of musicians in a semi-rural environment. Anderson has an ambivalent relationship with folk music and he also has an ambivalent attitude to the use of digital technologies. The importance of a second-hand digital delay pedal to his compositional processes stands out against his lack of interest in using dedicated digital sampling devices. This case study was about the *low-fidelity (or lo-fi)* use of sampling technologies, which continued a theme that was developed in earlier chapters of the book: the use of the Fairlight CMI Series I to create a 'grungy' sound and the lowering of sample rates by hip-hop producers like RZA to create a 'ghetto' sound.[2] The final case study was about a user who moves between the boundaries of different socio-musical worlds; the sampler is the instrument that allows him to do so. Matthew Herbert wants people to listen more carefully and pay closer attention to the sounds of the environment or what R. Murray Schafer referred to as 'the soundscape of the world' (1977, p. 3).[3] Herbert uses high-tech digital instruments but his work is part of a longer historical narrative that includes the modernist tradition of futurism (noise as music) and composers such as John Cage (everything as music). With a manifesto that supports his technological practices, this is sampling as *high art*.

Two other themes that were evident in the first half of the book also emerged during the writing up of its case studies: (i) accidents and (ii) authenticity. During the design of the Fairlight CMI, Peter Vogel accidentally discovered that sampling a sound was a more 'faithful' way of imitating acoustic instruments than using digital synthesis. As an example of serendipity in the recording studio, Marley Marl used an E-mu Emulator to sample drum sounds instead of a voice on a pre-existing recording. Accidents and mistakes formed part of Found's *unfettered sampling*: the sounds of falling rain or noises from next-door neighbours were left in their digital recordings and manipulated rather than removed.[4] As part of his lo-fi approach to music making, Kenny Anderson stressed that the use of a foot pedal to insert sampled sounds in his music was not the result of a conscious decision but something that he discovered by accident. Matthew Herbert made a deliberate choice to include mistakes and accidents in his music and this is part of his manifesto. Nick Prior writes that '[t]he history of technology and music are histories of misappropriation, accident, and contingency precisely because of the way objects are used and misused in practice' (2009, p. 86). Much of the material I have presented in the book supports this argument. The book also shows how users develop their own myths about misuse, mistakes, and the unconventional uses of digital technologies and why academics should be sensitive to ways musicians attempt to write themselves into these histories.

In the first half of the book I used the term *digital imperfection* to describe how users of the Fairlight CMI responded to its 'affordances' and sound fidelity issues. The use of accidental sounds mentioned in the previous paragraph might also be examples of digital imperfections, which are added to sound recordings to make them more authentic. This is the other theme that runs through the whole book: the importance of realism and authenticity to both the designers and users of digital sampling technologies. Claims about the high-fidelity levels of sampled sounds were central to the marketing campaigns of instrument designers at Fairlight Instruments, E-mu Systems, and New England Digital. Users of the Fairlight CMI such as Kate Bush were excited about recording the 'real sounds' of everyday life; composers at the BBC's Radiophonic Workshop welcomed being able to re-create the sounds of acoustic instruments using 'real sounds' rather than synthesized ones. The contemporary users of sampling technologies employed this same discourse of authenticity in each of my case studies. Leclair was critical of virtual instruments like software synthesizers that were unable to faithfully reproduce the sounds of hardware synthesizers: the Roland TB-303 was 'the real thing'. Members of Found used software samplers to imitate the sounds of acoustic instruments but preferred to use the 'real' instrument where possible as part of a more 'authentic' live performance. Anderson strived to use his sampler in 'uncontrived' ways and Herbert preferred to use 'real sounds' rather than pre-existing sounds. Edwards made no distinction between samplers and 'real' instruments, though he did express nostalgia for the analogue imperfections of pre-digitally recorded music – crackles, hisses, muffled sounds – that can now be digitally reproduced. This book has been about doing things in new ways with digital technologies. It has also showed the ongoing entanglement of analogue and digital technologies and how they continue to co-exist in the production of popular music.

One of the aims of this book has been to develop a conceptual framework for understanding the historical and contemporary uses of sampling instruments in a variety of socio-cultural contexts: in home studios, professional recording studios, on concert hall stages, and more mobile sites of musical production and performance. The instruments of contemporary music making have often been missing from the study of musical instruments and the technologies of digital sampling are often missing from the study of popular music. Synthesizer/sampling instruments like the Fairlight CMI and samplers like the Emulator are examples of the 'missing masses' of (popular) music studies. This book has tried to make them more visible. As the fields of organology and museology continue to change, those studying music technologies and the instruments of music making – acoustic, electric, analogue, *and* digital instruments – are working in interdisciplinary ways across a number of academic fields to shift the institutional boundaries that separate them.[5] By following the instruments, designers, users, and sellers of music technologies, we can try and make sense of the socio-technological processes of music making in the twentieth and early twenty-first century. By focusing on the *use* of these instruments – how musicians learn to use them, the failure to follow the instructions in manuals, the making of mistakes, the deliberate use of accidents, receiving reboot error messages, and the contingencies of musical practices – scholars can continue developing a more nuanced understanding about the relationship between humans and music making technologies.

Notes

1 Nick Prior writes that the laptop is 'an all-in-one production unit that meshes composition with dissemination and consumption. This is what differentiates the laptop from other mobile music devices such as the four-track portastudio, Walkman, or miniature keyboard. In effect, it is a meta-instrument, potentially containing all sounds (a feature it shares with the sampler) and production processes (a feature that transcends the sampler's capabilities)' (2008, p. 914).

2 Adam Harper defines lo-fi aesthetics as 'a positive appreciation of what are perceived and/or considered normatively interpreted as imperfections in a recording, with particular emphasis on imperfections in the recording technology itself' (2014, p. 6).

3 For more on soundscapes and the perceived problems of noise pollution in the late twentieth century, see Truax 1977. For discussions of the soundscape and its relevance to the fields of sound studies and anthropology, see Kelman 2010 and Samuels, Meintjes, Ochoa, and Porcello 2010.

4 For more on skipping CDs, malfunctioning electronics, and the use of accidental noises in the genre of glitch, see Cascone 2000, Young 2002, Bates 2004, Sangild 2004, and Kelly 2009.

5 See Pinch and Bijsterveld 2004 for their introduction to a special issue of *Social Studies of Science* on music technologies and the study of musical instruments as technological artifacts by scholars in ethnomusicology, history, anthropology, cultural studies, and sociology. For more on the interdisciplinary field of sound studies, see Sterne 2012a and Pinch and Bijsterveld 2012.

Interviews and Personal Communication

Anderson, K. 2009. Interview by the author. Minidisc recording. Scottish Fisheries Museum, Anstruther. 3 July.

Anderson, K. 2013. Private e-mail message. 21 February.

Burgess, R. 2011. Interview by the author. Minidisc recording. Skype. 21 May.

Campbell, Z. 2008. Interview by the author. Minidisc recording. Cameo Cinema Bar, Edinburgh. 11 October.

Cole, J. 2015. Private e-mail message. 18 September.

Coren, D. 2015. Interview by the author. E-mail. 17 June.

Cutler, C. 2008. Private e-mail message. 30 May.

Dean, P. 2015. Private e-mail message. 7 August.

Edwards, T. 2008. Interview by the author. E-mail. 23 June.

Herbert, M. 2012. Interview by the author. Minidisc recording. Summerhall, Edinburgh. 24 August.

Jeczalik, J. J. 2011. Interview by the author. Minidisc recording. Skype. 4 June.

Jones, C. 2015. Interview by the author. E-mail. 21 June.

Kelly, M. 2015. Interview by the author. E-mail. 21 September.

LeBlanc, K. 2008. Interview by the author. Minidisc recording. Voodoo Rooms, Edinburgh. 12 June.

Leclair, M. 2008a. Interview by the author. E-mail. 20 June.

Leclair, M. 2008b. Private e-mail message. 20 June.

Mendell, H. 2015. Interview by the author. E-mail. 19 June.

Perman, T. 2008. Interview by the author. Minidisc recording. Cameo Cinema Bar, Edinburgh. 11 October.

Rance, S. 2015. Interview by the author. E-mail. 31 August.

Rossum, D. 2018. Private e-mail message. 15 August.

Vogel, P. 2011a. Interview by the author. E-mail. 4 July.

Bibliography

Abildgaard, K. 2012. Interview with Dave Rossum. <http://theemus.com/interviews.html> Accessed 24 January 2019.

Acerra, M. 1983. Making Apple Music. *Creative Computing Buyer's Guide to Personal Computers, Peripherals, and Electronic Games.* pp. 147–150.

Aikin, J. 1983. Keyboard Report: The E-mu Drumulator. *Keyboard.* June. pp. 80–83.

Akai. 1985. Would you like a sampler™ for less than $1000.00?. *Keyboard.* July. p. 87.

Akrich, M. 1992. The De-Scription of Technical Objects. in W. Bijker and J. Law. eds. *Shaping Technology/Building Society: Studies in Sociotechnical Change.* Cambridge: The MIT Press. pp. 205–224.

Anderton, C. 1988a. Ensoniq Performance Sampler. *Sound on Sound.* February. pp. 36–38, 40.

Anderton, C. 1988b. Sampling Ensoniq. *Sound on Sound.* February. pp. 44–45, 47.

Anon. 1982. Kate Bush. *Electronics & Music Maker.* October. pp. 44–47.

Appleton, J. 1989. *21st-Century Musical Instruments: Hardware and Software.* New York: Institute for Studies in American Music.

Arrow, K. J. 1962. The Economic Implications of Learning by Doing. *The Review of Economic Studies.* 29:3. pp. 155–173.

Awde, N. 2008. *Mellotron: The Machine and the Musicians that Revolutionised Rock.* London: Desert Hearts.

Barr, T. 1998. *Kraftwerk: From Düsseldorf to the Future (with Love).* London: Ebury Press.

Barrett, C. 2008. Winter Wonderland. *Music Week.* 11 October. pp. 17–18.

Barry, H. V. 1987. Legal Aspects of Digital Sound Sampling. *Recording Engineer/Producer.* April. pp. 60–62, 67.

Bates, E. 2004. Glitches, Bugs, and Hisses: The Degeneration of Musical Recordings and the Contemporary Musical Work. in C. Washburne and M. Derno. eds. *Bad Music: The Music We Love to Hate.* New York: Routledge.

Batey, A. 1998. *Rhyming & Stealing: A History of the Beastie Boys.* London: Independent Music Press.

Beadle, J. J. 1993. *Will Pop Eat Itself? Pop Music in the Soundbite Era.* London: Faber and Faber.

Becker, H. 1974. Art as Collective Action. *American Sociological Review*. 39:6. pp. 767–776.

Becker, H. 1982 [2008]. *Art Worlds* (25th Anniversary Edition: Updated and Expanded). Berkeley: University of California Press.

Benson, R. 2000. UK Garage: The Secret History. *The Face*. June. pp. 56–63.

Bernstein, D. W. and Payne, M. 2008. Don Buchla. in D. W. Bernstein. ed. *The San Francisco Tape Music Center: 1960s Counterculture and the Avant-Garde*. Berkeley and Los Angeles: University of California Press.

Bijker W., Hughes, T., and Pinch, T. 1987 [2012]. eds. *The Social Construction of Technological Systems: New Directions in the Sociology and History of Technology*. Cambridge: The MIT Press.

Bijker, W. and Pinch, T. 2012. Preface to the Anniversary Edition. in W. Bijker, T. Hughes, and T. Pinch. eds. *The Social Construction of Technological Systems: New Directions in the Sociology and History of Technology*. Cambridge: The MIT Press. pp. xi–xxxiv.

Blake, A. 2007. *Popular Music: The Age of Multimedia*. London: Middlesex University Press.

Bohlman, P. 1999. Ontologies of Music. in N. Cook and M. Everist. eds. *Rethinking Music*. New York: Oxford University Press.

Boon, M. 2010. *In Praise of Copying*. Cambridge: Harvard University Press.

Born, G. and Hesmondhalgh, D. 2000. eds. *Western Music and its Others: Difference, Representation, and Appropriation in Music*. Berkeley and Los Angeles: University of California Press.

Bourdieu, P. 1984 [2010]. *Distinction: A Social Critique of the Judgement of Taste*. London and New York: Routledge.

Bourdieu, P. and Wacquant, L. 1992. *An Invitation to Reflexive Sociology*. Chicago: Chicago University Press.

Bourriaud, N. 2002. *Postproduction – Culture as Screenplay: How Art Reprograms the World*. New York: Lukas & Sternberg.

Boxer, S. 2015. The Synth Revival: Why the Moog is Back in Vogue. *The Guardian*. 8 July. <www.theguardian.com/music/2015/jul/08/the-grid-spearhead-synth-revival-moog-ensemble> Accessed 24 January 2019.

Brabazon, T. 2012. *Popular Music: Topics, Trends, & Trajectories*. London: Sage.

Bradby, B. 1993. Sampling Sexuality: Gender, Technology, and the Body in Dance Music. *Popular Music*. 12:2. pp. 155–176.

Brewster, B. and Broughton, F. 1999. *Last Night a DJ Saved My Life: The History of the Disc Jockey*. London: Headline.

Bright, S. 2000. *Peter Gabriel: An Authorised Biography*. London: Pan.

Briscoe, D. and Curtis-Bramwell, R. 1983. *The BBC Radiophonic Workshop: The First 25 Years*. London: BBC.

Bromberg, C. 1989. *The Wicked Ways of Malcolm McLaren*. London: Omnibus Press.

Brøvig-Hanssen, R. 2013. The Magnetic Tape Recorder: Recording Aesthetics in the New Era of Schizophonia. in T. Boon and F. Weium. eds. *Material Culture and Electronic Sound*. Washington: Smithsonian Institute Scholarly Press.

Brøvig-Hanssen, R. and Harkins, P. 2012. Contextual Incongruity and Musical Congruity: The Aesthetics and Humour of Mash-ups. *Popular Music*. 31:1. pp. 87–104

Brown, J. 2009. *Rick Rubin: In the Studio*. Toronto: ECW Press.

Burgess, R. J. 2005. *The Art of Music Production* (Third Edition). London: Omnibus Press.

Burgess, R. J. 2014. *The History of Music Production*. New York: Oxford University Press.

Buskin, R. 1994. From ABC to ZTT: The Amazing Career of Trevor Horn. *Sound on Sound*. August. pp. 34–41.

Buskin, R. 1995. The Sound that Anne Built. *Sound on Sound*. April. pp. 104–112.

Buskin, R. 2008. Classic Tracks: Afrika Bambaataa & The Soulsonic Force 'Planet Rock'. *Sound on Sound*. November. pp. 78–84.

Butler, M. 2006. *Unlocking the Groove: Rhythm, Meter, and Musical Design in Electronic Dance Music*. Bloomington and Indianapolis: Indiana University Press.

Campbell-Kelly, M. *et al.* 2018. *Computer: A History of the Information Machine* (Third Edition). New York: Routledge.

Carter, C. 1997. Happy Birthday MC!: Roland MC8 Microcomposer. *Sound on Sound*. March. pp. 280–282.

Cascone, K. 2000. The Aesthetics of Failure: 'Post-Digital' Tendencies in Contemporary Computer Music. *Computer Music Journal*. 24:4. pp. 12–18.

Chadabe, J. 1997. *Electric Sound: The Past and Promise of Electronic Music*. New Jersey: Prentice Hall.

Chang, J. 2005. *Can't Stop Won't Stop: A History of the Hip-Hop Generation*. London: Ebury Press.

Chapman, J. 2012. *Guide to the Qasar Tony Furse Archive*. Sydney: Powerhouse Museum.

Cloonan, M., Frith, S., and Williamson, J. 2003. *Mapping the Music Industry in Scotland: A Report*. Glasgow: Scottish Enterprise.

Coleman, B. 2007. *Check the Technique: Liner Notes for Hip-Hop Junkies*. New York: Villard Books.

Coleman, B. 2013. Davy DMX: Interview by Brian Coleman. in J. Mansfield. *Beat Box: A Drum Machine Obsession*. Malden: Get on Down.

Crombie, D. 1979. Man Bytes Dog. *Sound International*. December. p. 7.

Cunningham, M. 1998. *Good Vibrations: A History of Record Production* (Second Edition). London: Sanctuary.

Cutler, C. 1994. Plunderphonia. *Musicworks*. 60. pp. 6–19.

Danielsen, A. 2006. *Presence and Pleasure: The Funk Grooves of James Brown and Parliament*. Middletown: Wesleyan University Press.

Danielsen, A. 2010. Continuity and Break: James Brown's Funky Drummer. *PopScriptum*. 11.

Dawson, G. 1983. Machines alive with the Sound of Music. *New Scientist*. August. pp. 333–335.

Demers, J. 2006. *Steal This Music: How Intellectual Property Law Affects Musical Creativity*. Athens: University of Georgia Press.

Denyer, R. 1980. Eberhard Schoener. *Sound International*. May. pp. 14–16.

Dery, M. and Doerschuk, B. 1988. Drum Some Kill: The Beat Behind the Rap. *Keyboard*. November. pp. 34–36.

Diliberto, J. 1985. Kate Bush: From Piano to Fairlight with Britain's Exotic Chanteuse. *Keyboard*. July. pp. 56–73.

Doerschuk, B. 1983. The Great Synthesizer Debate. *Keyboard*. December. pp. 38–62.

Doerschuk, B. 1985. Many Mountains, Many Peaks: Encounters with the Driving Forces in Japan's Synthesizer Industry. *Keyboard*. August. pp. 48–58.

Elen, R. 1981. The CMI – An Insight into Digital Sound Synthesis. *Studio Sound*. February. pp. 44–46.

Elen, R. 1986. Fairlight Series III: The Ultimate Soft Machine. *Sound on Sound*. September. pp. 49–55.

E-mu. 1981. Play a Turkey. *Contemporary Keyboard*. May. p. 9.

E-mu. 1982. Breaking the Sound Barrier. *Keyboard*. October. p. 46.

E-mu. 1983a. Introducing the Drumulator from E-mu Systems, Inc. A Digital Drum Computer with an Amazing New Feature: Affordability. *Keyboard*. February. p. 37.

E-mu. 1983b. Drumulator Owner's Manual. Santa Cruz: E-mu Systems.

E-mu. 1985. E-mu Systems SP-12 Twelve-Bit Sampling Percussion System: The New Standard in Professional Digital Drum Computers. Santa Cruz: E-mu Systems.

E-mu. 1986. March to a Different Drummer. *Keyboard*. March. p. 71.

E-mu. 1987. Serious Repercussions: The SP-1200 Sampling Percussion System. *Electronic Musician*. October. p. 17.

E-mu. 1994. It's Baaaack. . .and It's Baaaad. Scotts Valley: E-mu Systems.

E-mu. 2015. E-mu: Birth of a Species. <www.creative.com/emu/company/history/> Accessed 24 January 2019.

Encarnacao, J. 2013. *Punk Aesthetics and New Folk: Way Down the Old Plank Road*. Aldershot: Ashgate.

Eno, B. 1983. The Studio as Compositional Tool – Part 1. *Down Beat*. July. pp. 56–57.

Eshun, K. 2000a. House. in P. Shapiro. ed. *Modulations: A History of Electronic Music*. New York: Caipirinha Productions.

Eshun, K. 2000b. House Rules. *The Wire*. December/January. pp. 37–39.

Evans, C. 1979 [1983]. *The Mighty Micro: The Impact of the Computer Revolution*. London: Victor Gollancz.

Fabbri, F. 2010. 'I'd Like my Record to Sound Like This': Peter Gabriel and Audio Technology. in M. Drewett, S. Hill, and K. Karki. eds. *Peter Gabriel: From Genesis to Growing Up*. Farnham: Ashgate.

Fairlight. 1980. Turn This Page and the Future of Music is Passed. . . *Sound International*. September. p. 5.

Fairlight. 1983a. *Fairlight Computer Musical Instrument*. Sydney: Fairlight Instruments.

Fairlight. 1983b. CMI – Extending Your Compositional Creativity. *Musician*. October. p. 80.

Fairlight. 1986. The Fairlight Series III – Specifications. Sydney: Fairlight Instruments.

Fairlight. 1987. How Much Fairlight Do you Need? *Studio Sound*. October. p. 67.

Fairlight. 2011a. *Fairlight CMI-30A: The Legend Returns*. Mona Vale: Fairlight Instruments.

Fairlight. 2011b. *Fairlight CMI App*. Mona Vale: Fairlight Instruments.

Falstrom, C. 1994. Thou Shalt Not Steal: *Grand Upright Music Ltd. v Warner Bros. Records, Inc.* and the Future of Digital Sound Sampling in Popular Music. *Hastings Law Journal*. 45. pp. 359–381.

Farber, J. 1980. Computers that Make Waves: Digital Synthesizers Take to the Road. *Rolling Stone*. 7 February. p. 64.

Feld, S. 2000. A Sweet Lullaby for World Music. *Public Culture*. 12:1. pp. 145–171.

Fink, R. 2005. The Story of ORCH5, or, The Classical Ghost in the Hip-Hop Machine. *Popular Music*. 24:3. pp. 339–356.

Fricke, J. and Ahearn, C. 2002. *Yes Yes Y'All: Oral History of Hip-Hop's First Decade*. Oxford: The Perseus Press.

Frith, S. 1981. 'The Magic That Can Set You Free': The Ideology of Folk and the Myth of the Rock Community. *Popular Music*. 1. pp. 159–168.

Frith, S. 1986. Art Versus Technology: The Strange Case of Popular Music. *Media, Culture, and Society*. 8:3. pp. 263–279.

Frith, S. 1992. The Industrialisation of Music. in J. Lull. ed. *Popular Music and Communication*. London: Sage.

Frith, S. 1993. Music and Morality. in S. Frith. ed. *Music and Copyright*. Edinburgh: Edinburgh University Press.

Frith, S. 1996. *Performing Rites: On the Value of Popular Music*. Oxford: Oxford University Press.

Frith, S. 2004. What is Bad Music? in C. Washburne and M. Derno. eds. *Bad Music: The Music We Love to Hate*. New York: Routledge.

Frith, S. and Horne, H. 1987. *Art into Pop*. London: Metheun.

Galloway, V. 2013. *Songs in the Key of Fife: The Intertwining Stories of The Beta Band, King Creosote, KT Tunstall, James Yorkston, and the Fence Collective*. Edinburgh: Polygon.

Gibson, J. 1979. *The Ecological Approach to Visual Perception*. Boston: Houghton Mifflin Company.

Gilby, P. 1985. Sampling the Japanese Way. *Sound on Sound*. December. pp. 34–38.

Gilby, P. 1987a. Kim Ryrie: Fairlight Instruments. *Sound on Sound*. May. pp. 50–52.

Gilby, P. 1987b. The Sound on Sound Guide to Samplers. *Sound on Sound*. November. pp. 34–40.

Goldmann, S. 2015. *Presets – Digital Shortcuts to Sound*. London: The Bookworm.

Goodwin, A. 1988. Sample and Hold: Pop Music in the Digital Age of Reproduction. *Critical Quarterly*. 30:3. pp. 34–49.

Grandl, P. 2015a. From Biology to Modular System – Interview: Dave Rossum E-mu, Part One. *Amazona Musiker Magazin*. 2 July. <www.amazona.de/interview-dave-rossum-e-mu-part-one-english-version> Accessed 24 January 2019.

Grandl, P. 2015b. Golden Years, Whap Whap Whap – Interview: Dave Rossum E-mu, Part Two. *Amazona Musiker Magazin*. 3 July. <www.amazona.de/interview-dave-rossum-e-mu-part-two-english-version/4/> Accessed 24 January 2019.

Grandl, P. 2015c. From Emax to Proteus – Interview: Dave Rossum E-mu, Part Three. *Amazona Musiker Magazin*. 4 July. <www.amazona.de/interview-dave-rossum-e-mu-teil-3/> Accessed 24 January 2019.

Gray, L. 1987. Fairlight Robbery. *New Musical Express*. 11 July. pp. 28–29, 37.

Greenwald, T. and Burger, J. 2000. It Came from the Music Industry. in M. Vail. ed. *Vintage Synthesizers*. San Francisco: Miller Freeman. pp. 82–102.

Gregory, G. 1985. *Japanese Electronics Technology: Enterprise and Innovation.* New York: John Wiley.

Gregory, W. 2015. Back to the Future: I'm in the Moog Again. *The Guardian.* 9 June <www.theguardian.com/music/musicblog/2015/jun/09/moog-synthesiser-revival-70s-monosynths-will-gregory-moog-ensemble> Accessed 24 January 2019.

Hamer, M. 2005. Electronic Maestros. *New Scientist.* 26 March. pp. 48–51.

Hammond, R. 1983. *The Musician and the Micro.* Dorset: Blandford Press.

Hansen, K. F. 2002. The Basics of Scratching. *Journal of New Music Research.* 31:4. pp. 357–365.

Hansen, K. F. 2015. DJs and Turntablism. in J. Williams. ed. *The Cambridge Companion to Hip-Hop.* Cambridge: Cambridge University Press.

Harper, A. 2014. Lo-Fi Aesthetics in Popular Music Discourse. University of Oxford: PhD Thesis.

Hastings, T. 1986a. Making the Most of Your Mirage. *Sound on Sound.* October. pp. 12–13.

Hastings, T. 1986b. Making the Most of Your Mirage. *Sound on Sound.* November. pp. 29–30.

Haynes, J. 2001. Doctors of Madness. *The Wire.* April. pp. 26–28.

Herbert, M. 2005a. Personal Contract for the Composition of Music [Incorporating the Manifesto of Mistakes]. <www.matthewherbert.com/manifesto/> Accessed 24 January 2019.

Herbert, M. 2005b. Liner note to *Plat du Jour.* Compact Disc. Accidental Records.

Herbert, M. 2011. Liner note to *One Pig.* Compact Disc. Accidental Records.

Herbert, M. 2013a. Sonography. <www.matthewherbert.com/sonography> Accessed 24 January 2019.

Herbert, M. 2013b. Liner note to *The End of Silence.* Compact Disc. Accidental Records.

Herrmann, T. 2002. Akufen: Electronic Music and DJ. *Ableton.* <www.ableton.com/pages/artists/akufen> Accessed 24 January 2019.

Hesmondhalgh, D. 2006. Digital Sampling and Cultural Inequality. *Social & Legal Studies.* 15:1. pp. 53–75.

Holmes, T. 2016. *Electronic and Experimental Music: Technology, Music, and Culture* (Fifth Edition). London: Routledge.

Holm-Hudson, K. 1996. John Oswald's *Rubaiyat (Elektrax)* and the Politics of Recombinant Do-Re-Mi. *Popular Music and Society.* 20:3. pp. 19–36.

Holm-Hudson, K. 1997. Quotation and Context: Sampling and John Oswald's Plunderphonics. *Leonardo Music Journal.* 7. pp. 17–25.

Hoskyns, B. 1984. How to Make a Spectacle of Yourself. *New Musical Express.* 13 October. pp. 24–26.

Host, V. 2002. It's a Spiritual Thing. *Deuce Magazine.* November. pp. 18–21.

Hsu, H. 2008. Liner note to *What Does It All Mean? 1983–2006 Retrospective.* Compact Disc. Illegal Art.

Hunter, D. 2013. *Guitar Effects Pedals: The Practical Handbook.* London: Outline Press.

Husband, S. 1985. Art of Noise: Raiders of the Twentieth Century. *No. 1.* 23 February. pp. 19–20.

Hutchby, I. 2001. Technologies, Texts, and Affordances. *Sociology*. 35:2. pp. 441–456.

Igma, N. 1990a. Recipes for Plunderphonic: An Interview with John Oswald, Part 1. *Musicworks*. 47. pp. 4–10.

Igma, N. 1990b. Taking Sampling 50 Times Beyond the Expected: An Interview with John Oswald, Part 2. *Musicworks*. 48. pp. 16–21.

Ingram, M. 2009. Switched-On. *Loops*. Issue 1. pp. 129–137.

Jenkins, M. 1987. Casio FZ-1 Sampling Keyboard: Leading the Field?. *Sound on Sound*. June. pp. 62–66.

Jenkins, M. 2007. *Analog Synthesizers: Understanding, Performing, Buying*. New York: Focal Press.

Kakehashi, I. 2002. *I Believe in Music: Life Experiences and Thoughts on the Future of Electronic Music by the Founder of the Roland Corporation*. Milwaukee: Hal Leonard.

Kärjä, A. 2006. A Prescribed Alternative Mainstream: Popular Music and Canon Formation. *Popular Music*. 25:1. pp. 3–19.

Katz, M. 2004. *Capturing Sound: How Technology has Changed Music*. Los Angeles: University of California Press.

Katz, M. 2006. Men, Women, and Turntables: Gender and the DJ Battle. *The Musical Quarterly*. 89:4. pp. 580–599.

Katz, M. 2012. *Groove Music: The Art and Culture of the Hip-Hop DJ*. New York: Oxford University Press.

Kawohl, F. and Kretschmer, M. 2009. Johann Gottlieb Fichte and the Trap of *Inhalt* (Content) and Form. *Information, Communication, and Society*. 12:2. pp. 205–228.

Keeble, R. 2002. 30 Years of E-mu: The History of E-mu Systems. *Sound on Sound*. September. pp. 121–127.

Kellner, C. Lapham, E., and Spiegel, L. 1980. The Alphasyntauri: A Keyboard Based Digital Playing and Recording System with a Microcomputer Interface. Paper presented at the 67th Audio Engineering Society convention. New York. 1 November.

Kelly, C. 2009. *Cracked Media: The Sound of Malfunction*. Cambridge: The MIT Press.

Kelman, A. 2010. Rethinking the Soundscape: A Critical Genealogy of a Key Term in Sound Studies. *Senses & Society*. 5:2. pp. 212–234.

Kemp, M. 1992. The Death of Sampling?. *Option*. March/April. pp. 17–20.

Keyboard Staff. 1983. Herbie Hancock. *Keyboard*. February. pp. 47–59.

Kirn, P. 2011. Robert Henke (Monolake): The Composer, Artist, and Ableton Live Imagineer Looks to the Future. in P. Kirn. ed. *The Evolution of Electronic Dance Music*. Milwaukee: Backbeat Books.

Laderman, D. and Westrup, L. 2014. eds. *Sampling Media*. Oxford: Oxford University Press.

Lane, C. and Carlyle, A. 2013. *In the Field: The Art of Field Recording*. Axminster: Uniformbooks.

Latham, S. 2003. *Newton v Diamond*: Measuring the Legitimacy of Unauthorised Compositional Sampling – A Clue Illuminated and Obscured. *Hastings Communications and Entertainment Law Journal*. 26. pp. 119–153.

Leclair, M. 2001. Liner note to *My Way*. Compact disc. Force Inc.

Lee, J. 1981. Interview: Dave Rossum. *Polyphony*. November/December. pp. 23–35.

Leete, N. 1999. The Fun of the Fairlight: Fairlight Computer Musical Instrument (Retro). *Sound on Sound*. April. pp. 254–259.

Lehrman, P. 1983. New Technologies. *High Fidelity*. September. pp. 51–54.

Levine, S. and Mauchly, J. W. 1980. The Fairlight Computer Music Instrument. in *Proceedings of the 1980 International Computer Music Conference*. pp. 565–573.

Levine, S. and Mauchly, B. 1981. Hardware Review: alphaSyntauri Music Synthesizer. *Byte*. December. pp. 108–128.

Linn. n.d. Real Drums at Your Fingertips. Hollywood: Linn Electronics, Inc.

Linn. 1982. LinnDrum: The Ultimate Drum Machine. *Keyboard*. April. p. 13.

Loza, S. 2001. Sampling (Hetero)Sexuality: Diva-ness and Discipline in Electronic Dance Music. *Popular Music*. 20:3. pp. 349–357.

M3 Event. 2012. Interview – Venetian Snares. 14 March <www.m3event.wordpress.com/2012/03/14/interview-venetian-snares/> Accessed 24 January 2019.

MacKenzie, D. and Wajcman, J. 1985 [2011]. Introductory Essay: The Social Shaping of Technology. in D. MacKenzie and J. Wajcman. eds. *The Social Shaping of Technology* (Second Edition). Maidenhead: Open University Press.

Malsky, M. 2003. Stretched from Manhattan's Back Alley to MOMA: A Social History of Magnetic Tape Recording. in R. Lysloff and L. Gay Jr. eds. *Music and Technoculture*. Middletown: Wesleyan University Press.

Manning, P. 2013. *Electronic and Computer Music* (Fourth Edition). New York: Oxford University Press.

Mansfield, J. 2013. *Beat Box: A Drum Machine Obsession*. Malden: Get on Down.

Marcus, J. H. 1991. Don't Stop That Funky Beat: The Essentiality of Digital Sampling to Rap Music. *Hastings Communications and Entertainment Law Journal*. 13. pp. 767–790.

Marshall, W. 2006. Giving Up Hip-Hop's Firstborn: A Quest for the Real after the Death of Sampling. *Callaloo*. 29:3. pp. 1–25.

Martin, P. 1984. Who or What is The Art of Noise? *Smash Hits*. 25 October–7 November. pp. 34–35.

Mason, A. 2008. Pete Rock Reminisces. *Wax Poetics Anthology Volume 2*. New York: Wax Poetics Books.

Mathews, M. 1963. The Digital Computer as a Musical Instrument. *Science*. 142:3592. pp. 553–557.

Mathews, M. 1969. *The Technology of Computer Music*. Cambridge: The M.I.T. Press.

Matos, M. 2003. Ransom Notes To God: Information Overload You Can Dance To. *The Village Voice*. 22 April. <www.villagevoice.com/2003/04/22/ransom-notes-to-god/> Accessed 24 January 2019.

Matos, M. 2007. Todd Edwards: The Stylus Interview. *Stylus*. 23 April. <www.stylusmagazine.com/articles/weekly_article/todd-edwards-the-stylus-interview.htm> Accessed 2 February 2016.

Maxwell, T. 2009. Anderson's Not Sitting on the Fence with His Latest Album. *Edinburgh Evening News*. 22 May. p. 5.

McLeod, K. 2004. How Copyright Law Changed Hip-Hop: An Interview with Public Enemy's Chuck D and Hank Shocklee about Hip-Hop, Sampling, and How Copyright Law Altered the Way Hip-Hop Artists Made Their Music. *Stay Free! Magazine*. May.

McLeod, K. 2007. *Freedom of Expression: Resistance and Repression in the Age of Intellectual Property*. Minneapolis: University of Minnesota Press.

McLeod, K. and DiCola, P. 2011. *Creative License: The Law and Culture of Digital Sampling*. Durham: Duke University Press.

McLeod, K. and Kuenzli, R. 2011. eds. *Cutting Across Media: Appropriation Art, Interventionist Collage, and Copyright Law*. Durham: Duke University Press.

Meintjes, L. 1990. Paul Simon's *Graceland*, South Africa, and the Mediation of Musical Meaning. *Ethnomusicology*. 34:1. pp. 37–73.

Mellor, D. 1987. E-mu Systems SP-1200 Sampling Percussion System. *Sound on Sound*. October. pp. 22–28.

Meredith, H. 1981. Computer Music in/Concert in High Places. *Pacific Computer Weekly*. 30 October-5 November. pp. 7, 14.

Mico, T. 1985. Is Anybody Still Afraid of The Art of Noise? *Melody Maker*. 19 October. pp. 14–15.

Miller, P. D. 2008. ed. *Sound Unbound: Sampling Digital Music and Culture*. Cambridge: The MIT Press.

Milner, G. 2009. *Perfecting Sound Forever: The Story of Recorded Music*. London: Granta.

Moog, B. 1981. On Synthesizers: The Alpha Syntauri. *Keyboard*. November. pp. 76–77.

Moog, B. 1983. M.I.D.I. (Musical Instrument Digital Interface): What it is, What it Means to You. *Keyboard*. July. pp. 19–25.

Moog, B. 1985. The Keyboard Explosion: Ten Amazing Years in Music Technology. *Keyboard*. October. pp. 36–48.

Moog, B. 1989. You've Come a Long Way, MIDI. *Keyboard*. February. pp. 117, 168.

Morey, J. 2007. The Death of Sampling: Has Litigation Destroyed an Art Form? Proceedings of the 3rd Art of Record Production Conference. Queensland University of Technology. 10–11 December.

Morey, J. 2012. The Bridgeport Dimension: Copyright Enforcement and its Implications for Sampling Practice. in A. Kärjä, L. Marshall, and J. Brusila. eds. *Music, Business, and Law: Essays on Contemporary Trends in the Music Industry*. Helsinki: International Institute for Popular Culture. pp. 21–45.

Mueller, J. 2006. All Mixed Up: Bridgeport Music v. Dimension Films and De Minimis Digital Sampling. *Indiana Law Journal*. 81:1. pp. 435–463.

Myrie, R. 2008. *Don't Rhyme for the Sake of Riddlin': The Authorised Story of Public Enemy*. Edinburgh: Canongate.

Nelson, H. 1991. Soul Controller: Sole Survivor. *The Source*. October. pp. 37–39.

Niebur, L. 2010. *Special Sound: The Creation and Legacy of the BBC Radiophonic Workshop*. New York: Oxford University Press.

Oberheim. 1981. Oberheim Brings the Studio to the Stage. *Keyboard*. November. pp. 46–47.

Ochoa Gautier, A. M. 2006. Sonic Transculturation, Epistemologies of Purification, and the Aural Public Sphere in Latin America. *Social Identities*. 12:6. pp. 803–825.

Olson, H. F. and Belar, H. 1955. Electronic Music Synthesizer. *The Journal of the Acoustical Society of America*. 27:3. pp. 595–612.

O'Neil, T. 2006. A Thousand Different Keys: A Lunchtime Conversation with Matthew Herbert. *PopMatters*. 7 July. <www.popmatters.com/pm/feature/herbert-matthew-060707> Accessed 28 August 2013.

Oppenheimer, L. 1986. The E-mu SP-12. *Electronic Musician*. July. pp. 84–90.

Oswald, J. 1986. Plunderphonics or Audio Piracy as a Compositional Prerogative. *Musicworks*. 34. pp. 5–8.

Oswald, J. 1988. Neither a Borrower nor a Sampler Prosecute. *Keyboard*. March. pp. 12, 14.

Oudshoorn, N. and Pinch, T. 2003 [2005]. eds. *How Users Matter: The Co-Construction of Users and Technology*. Cambridge: The MIT Press.

Peel, I. 2005. Trevor Horn: 25 Years of Hits. *Sound on Sound*. March. pp. 50–57.

Peters, B. 2016. Digital. in B. Peters. ed. *Digital Keywords: A Vocabulary of Information Society and Culture*. New Jersey: Princeton University Press.

Pinch, T. 2003 [2005]. Giving Birth to New Users: How the Minimoog was Sold to Rock and Roll. in N. Oudshoorn and T. Pinch. eds. *How Users Matter: The Co-Construction of Users and Technology*. Cambridge: The MIT Press. pp. 247–270.

Pinch, T. and Bijker, W. 1984. The Social Construction of Facts and Artefacts: Or How the Sociology of Science and the Sociology of Technology Might Benefit Each Other. *Social Studies of Science*. 14:3. pp. 399–441.

Pinch, T. and Bijsterveld, K. 2004. Sound Studies: New Technologies and Music. *Social Studies of Science*. 34:5. pp. 635–648.

Pinch, T. and Bijsterveld, K. 2012. New Keys to the World of Sound. in T. Pinch and K. Bijsterveld. eds. *The Oxford Handbook of Sound Studies*. New York: Oxford University Press.

Pinch, T. and Trocco, F. 2002. *Analog Days: The Invention and Impact of the Moog Synthesizer*. Cambridge: Harvard University Press.

Porcello, T. 1991. The Ethics of Digital Audio-Sampling: Engineers' Discourse. *Popular Music*. 10:1. pp. 69–84.

Poschardt, U. 1995 [1998]. *DJ Culture*. London: Quartet Press.

Pouncey, E. 2002. Rock Concrète: Counterculture Plugs in to the Academy. in R. Young. ed. *Undercurrents: The Hidden Wiring of Modern Music*. London: Continuum.

Pressing, J. 1992. *Synthesizer Performance and Real-Time Techniques*. Oxford: Oxford University Press.

Prior, N. 2007. The Socio-technical Biography of a Musical Instrument: The Case of the Roland TB-303. Paper delivered at BSA Annual Conference. University of East London. April.

Prior, N. 2008. OK Computer: Mobility, Software, and the Laptop Musician. *Information, Communication, & Society*. 11:7. pp. 912–932.

Prior, N. 2009. Software Sequencers and Cyborg Singers: Popular Music in the Digital Hypermodern. *New Formations*. 66. pp. 81–99.

Prior, N. 2017. On Vocal Assemblages: From Edison to Miku. *Contemporary Music Review*. pp. 1–19.

Prior, N. 2018. *Popular Music, Digital Technology, and Society*. London: Sage.

Read, P. 1999. Todd Edwards. *DJ Magazine*. 9–22 October. pp. 54–55.

Read, P. 2001. Todd Edwards. *Seven Magazine*. 27 June. pp. 25–26.

Reid, G. 2002. Rebirth of the Cool: Mellotron. *Sound on Sound*. August. pp. 204–211.

Reid, G. 2004a. Designing the Future: The History of Roland Part 1 – 1930–1978. *Sound on Sound*. November. pp. 107–121.

Reid, G. 2004b. Designing the Future: The History of Roland Part 2 – 1979–1985. *Sound on Sound*. December. pp. 129–145.

Reid, G. 2005a. Designing the Future: The History of Roland Part 3 – 1986–1991. *Sound on Sound*. January. pp. 139–152.

Reid, G. 2005b. Designing the Future: The History of Roland Part 4 – 1992–1997. *Sound on Sound*. February. pp. 117–131.

Reynolds, S. 1999. Adult Hardcore. *The Wire*. April. pp. 54–58.

Reynolds, S. 2003. Fave Albums of 2002. *Blissblog*. 5 January. <www.blissout.blogspot.com/2003_01_01_archive.html#86993298> Accessed 24 January 2019.

Reynolds, S. 2005. *Rip it Up and Start Again: Postpunk 1978–1984*. London: Faber and Faber.

Reynolds, S. and Stubbs, D. 1990. Sampling. in S. Reynolds. *Blissed Out: The Raptures of Rock*. London: Serpent's Tail.

Richardson, J. 2009. Televised Live Performance, Looping Technology, and the 'Nu Folk': KT Tunstall on *Later. . .with Jools Holland*. in D. B. Scott. ed. *The Ashgate Research Companion to Popular Musicology*. Farnham: Ashgate.

Robertson, M. 2006. Found: Pick of 2006 – Hot 100. 12 December. *The List*. <www.list.co.uk/article/13059-pick-of-2006-hot-100/> Accessed 24 January 2019.

Robertson, S. 1983. Malcolm: 'Soweto' Plundered? *New Musical Express*. 2 April. p. 3.

Rodgers, N. 2011. *Le Freak: An Upside Down Story of Family, Disco, and Destiny*. London: Sphere.

Rodgers, T. 2003. On the Process and Aesthetics of Sampling in Electronic Music Production. *Organised Sound*. 8:3. pp. 313–320.

Rodgers, T. 2010. Introduction. *Pink Noises: Women on Electronic Music and Sound*. Durham: Duke University Press.

Rodgers, T. 2015. Tinkering with Cultural Memory: Gender and the Politics of Synthesizer Historiography. *Feminist Media Histories*. 1:4. pp. 5–30.

Rognoni, G. R. 2017. Organology and the Others: A Political Perspective. Discussion paper prepared for The Galpin Society and American Musical Instrument Society joint conference. The University of Edinburgh. 3 June.

Roman, S. 2001. *Digital Babylon: Hollywood, Indiewood, and Dogme 95*. Hollywood: iFilm Publishing.

Rose, T. 1994. *Black Noise: Rap Music and Black Culture in Contemporary America*. Hanover: Wesleyan University Press.

Rosenberg, N. 1982. *Inside the Black Box: Technology and Economics*. Cambridge: Cambridge University Press.

RZA, The. 2005. *The Wu-Tang Manual*. London: Plexus Publishing.

Saiber, A. 2007. The Polyvalent Discourse of Electronic Music. *PMLA*. 122:5. pp. 1613–1625.

Samagaio, F. 2002. *The Mellotron Book*. Vallejo: ProMusic Press.

Samuels, D., Meintjes, L., Ochoa, A. M., and Porcello, T. 2010. Soundscapes: Towards a Sounded Anthropology. *Annual Review of Anthropology*. 39. pp. 329–345.

Sanctuary. 2005. The Sugarhill Story. Liner note to *The Message: The Story of Sugarhill Records*. Compact Disc. Sanctuary.

Sanden, P. 2012. Virtual Liveness and Sounding Cyborgs: John Oswald's 'Vane'. *Popular Music*. 31:1. pp. 45–68.

Sandywell, B. 2004. The Myth of the Everyday. *Cultural Studies*. 18:2/3. pp. 160–180.

Sangild, T. 2004. Glitch – The Beauty of Malfunction. in C. Washburne and M. Derno. eds. *Bad Music: The Music We Love to Hate*. New York: Routledge.

Sanjek, D. 1994. 'Don't Have to DJ No More': Sampling and the 'Autonomous' Creator. in M. Woodmansee and P. Jaszi. eds. *The Construction of Authorship: Textual Appropriation in Law and Literature*. Durham: Duke University Press.

Scannell, J. 2012. *James Brown*. Sheffield: Equinox Publishing.

Scarth, G. 2013. Interviews: Roger Linn on Swing, Groove, and the Magic of the MPC's Timing. *Attack*. <www.attackmagazine.com/features/interview/roger-linn-swing-groove-magic-mpc-timing/3/> Accessed 24 January 2019.

Schafer, R. M. 1977 [1994]. *The Soundscape: Our Sonic Environment and the Tuning of the World*. Rochester: Destiny Books.

Schietinger, J. 2005. Bridgeport Music, Inc. v. Dimension Films: How the Sixth Circuit Missed a Beat on Digital Music Sampling. *DePaul Law Review*. 55:1. pp. 209–248.

Schloss, J. G. 2014 [2004]. *Making Beats: The Art of Sample-Based Hip-Hop*. Conneticut: Wesleyan University Press.

Schumacher, T. G. 1995. 'This is a Sampling Sport': Digital Sampling, Rap Music, and the Law in Cultural Production. *Media, Culture, & Society*. 17:2. pp. 253–273.

Sequential Circuits. 1984a. Your Personal Orchestra. *Keyboard*. February. pp. 50–51.

Sequential Circuits. 1984b. Your Personal Orchestra. *Electronics & Music Maker*. July. pp. 52–53.

Shapiro, P. 2002. Deck Wreckers: The Turntable as Instrument. in R. Young. ed. *Undercurrents: The Hidden Wiring of Modern Music*. London: Continuum. pp. 163–176.

Sherburne, P. 2001. The Rules of Reduction. *The Wire*. July. pp. 18–25.

Shuker, R. 2017. *Popular Music: The Key Concepts* (Fourth Edition). London: Routledge.

Simons, J. 2003. *Playing the Waves: Lars von Trier's Game Cinema*. Amsterdam: Amsterdam University Press.

Smith, S. 2013. *Hip-Hop Turntablism, Creativity, and Collaboration*. Farnham: Ashgate.

St. Michael, M. 1994. *Peter Gabriel: In His Own Words*. London: Omnibus Press.

Star, S. L. and Griesemer, J. R. 1989. Institutional Ecology, 'Translations', and Boundary Objects: Amateurs and Professionals in Berkeley's Museum of Vertebrate Zoology, 1907–39. *Social Studies of Science*. 19. pp. 387–420.

Steenhuisen, P. 2009. John Oswald. in *Sonic Mosaics: Conversations with Composers*. Edmonton: University of Alberta Press.

Sterne, J. 2003. *The Audible Past: Cultural Origins of Sound Reproduction*. Durham: Duke University Press.

Sterne, J. 2006. What's Digital in Digital Music? in P. Messaris and L. Humphreys. eds. *Digital Media: Transformations in Human Communication*. New York: Peter Lang.

Sterne, J. 2012a. ed. *The Sound Studies Reader*. London and New York: Routledge.

Sterne, J. 2012b. *MP3: The Meaning of a Format*. Durham: Duke University Press.

Sterne, J. 2016. Analog. in B. Peters. ed. *Digital Keywords: A Vocabulary of Information Society and Culture*. New Jersey: Princeton University Press.

Stewart, A. 2000. 'Funky Drummer': New Orleans, James Brown, and the Rhythmic Transformation of American Popular Music. *Popular Music*. 19:3. pp. 293–318.

Stewart, A. 2005. The Name Behind the Name: Bruce Jackson – Apogee, Jands, Lake Technology. *Audio Technology*. May. pp. 65–70.

Stock, J. 1995. Reconsidering the Past: Zhou Xuan and the Rehabilitation of Early Twentieth-Century Popular Music. *Asian Music*. 26:2. pp. 119–135.

Strachan, R. 2007. Micro-independent Record Labels in the UK: Discourse, DIY Cultural Production, and the Music Industry. *European Journal of Cultural Studies*. 10:2. pp. 245–265.

Street, R. 2000. Fairlight: A 25-Year Long Fairytale. *Audio Media*. 8 November.

Street, J., Zhang, L., Simuniak, M., and Wang, Q. 2015. *Copyright and Music Policy in China: A Literature Review*. CREATe Working Paper 2015/16 (August 2015).

Stubbs, D. 2012. Biography. <www.matthewherbert.com/biography> Accessed 23 January 2013.

Sutcliffe, P. 1987. Sound Wars. *Q*. December. pp. 10–12.

Syco Systems. 1981. From Farts to Filharmonics: The Emulator – New System from Syco. *Studio Sound*. August. pp. 4–5.

Synclavier. 1980. Announcing the End of Synthesizers as You Now Know Them. *Musician, Player, and Listener*. November. pp. 50–51.

Taylor, P. 1988. *Impresario: Malcolm McLaren and the British New Wave*. New York: New Museum of Contemporary Art.

Taylor, T. D. 2001. *Strange Sounds: Music, Technology, & Culture*. London: Routledge.

Taylor, T. D. 2003. A Riddle Wrapped in a Mystery: Transnational Music Sampling and Enigma's 'Return to Innocence'. in R. Lysloff and L. Gay Jr. eds. *Music and Technoculture*. Middletown: Wesleyan University Press.

Taylor, T. D. 2007. *Beyond Exoticism: Western Music and the World*. Durham: Duke University Press.

Théberge, P. 1997. *Any Sound You Can Imagine: Making Music/Consuming Technology*. Middletown: Wesleyan University Press.

Théberge, P. 2003. 'Ethnic Sounds': The Economy and Discourse of World Music Sampling. in R. Lysloff and L. Gay Jr. eds. *Music and Technoculture*. Middletown: Wesleyan University Press.

Théberge, P. 2004. Technology, Creative Practice, and Copyright. in S. Frith and L. Marshall. eds. *Music and Copyright* (Second Edition). Edinburgh: Edinburgh University Press.

Théberge, P. 2015. Digitalisation. in J. Shepherd and K. Devine. eds. *The Routledge Reader on the Sociology of Music*. London: Routledge.

Thompson, E. 2005. Machines, Music, and the Quest for Fidelity: Marketing the Edison Phonograph in America, 1877–1925. *The Musical Quarterly.* 79:1. pp. 131–171.

Thomson, G. 2010. *Under the Ivy: The Life and Music of Kate Bush.* London: Omnibus Press.

Tingen, P. 1993. Ratman Returns: JJ Jeczalik and The Art of Toys. *Recording Musician.* January. pp. 52–56.

Tingen, P. 1996a. Fairlight: The Whole Story. *Audio Media.* January. pp. 48–55.

Tingen, P. 1996b. Noises Off: JJ Jeczalik – Art of Silence. *Sound on Sound.* August. pp. 98–105.

Toop, D. 1999. *Exotica: Fabricated Soundscapes in a Real World.* London: Serpent's Tail.

Toop, D. 2000. Hip-Hop: Iron Needles of Death and a Piece of Wax. in P. Shapiro. ed. *Modulations: A History of Electronic Music.* New York: Caipirinha Productions.

Torchia, D. 1987. Sampling Realities: Frank Zappa's Experience with his Recent *Jazz From Hell* Album. *Recording Engineer/Producer.* April. p. 64.

Tresch, J. and Dolan, E. I. 2013. Toward a New Organology: Instruments of Music and Science. *Osiris.* 28:1. pp. 278–298.

Truax, B. 1977. The Soundscape and Technology. *Interface.* 6. pp. 1–8.

Vaidhyanathan, S. 2001. *Copyrights and Copywrongs: The Rise of Intellectual Property and How It Threatens Creativity.* New York: New York University Press.

Vail, M. 1990. Vintage Synths: The Roland MC-8 MicroComposer. *Keyboard.* October. pp. 116–117.

Vail, M. 1994. Vintage Synths: Roland CR-78, TR-808, & TR-909. *Keyboard.* May. pp. 82–86.

Vail, M. 2000a. EMS VCS3 & Synthi A/AKS. in M. Vail. ed. *Vintage Synthesizers.* San Francisco: Miller Freeman. pp. 110–114.

Vail, M. 2000b. Fairlight CMI: Trailblazing Megabuck Sampler. in M. Vail. ed. *Vintage Synthesizers.* San Francisco: Miller Freeman. pp. 214–219.

Vail, M. 2000c. E-mu Emulator: First Affordable Digital Sampler. in M. Vail. ed. *Vintage Synthesizers.* San Francisco: Miller Freeman. pp. 220–225.

Vail, M. 2000d. Mellotron: Pillar of a Musical Genre. in M. Vail. ed. *Vintage Synthesizers.* San Francisco: Miller Freeman. pp. 229–235.

Vail, M. 2014. *The Synthesizer: A Comprehensive Guide to Understanding, Programming, Playing, and Recording the Ultimate Music Instrument.* New York: Oxford University Press.

Vogel, P. 2011b. The ORCH2/5 Story. *The Peter Vogel Instruments Forum.* 20 June.

Waksman, S. 1999. *Instruments of Desire: The Electric Guitar and the Shaping of Musical Experience.* Cambridge: Harvard University Press.

Walker, M. 1998. Nemesys Gigasampler v1.5: PC Software Sampler. *Sound on Sound.* December. <www.soundonsound.com/sos/dec98/articles/gigasample.143.htm> Accessed 9 April 2016.

Warner, T. 2003. *Pop Music – Technology and Creativity: Trevor Horn and the Digital Revolution.* Aldershot: Ashgate.

Webley, G. 1998a. Paid to Play: All About Session Musicians. *Sound on Sound*. October. pp. 40–46.

Webley, G. 1998b. Paid to Play: A Personal View of Session Programming. *Sound on Sound*. November. pp. 28–32.

Whelan, A. 2009. The 'Amen' Breakbeat as Fratriarchal Totem. in B. Neumeier. ed. *Dichotonies: Gender and Music*. Heidelberg: Universitätsverlag Winter.

White, P. 1995. Digidesign Past & Present: Interview with the President. *Sound on Sound*. March. pp. 36–41.

White, P. 1999. Fresh Cream: Akai S5000 and S6000 Samplers. *Sound on Sound*. January. pp. 88–95.

Wiffen, P. 1985. Linn 9000: Digital Drum Machine and Keyboard Recorder. *Electronics and Music Maker*. April. pp. 26–30.

Wiffen, P. 1988. Emulator III: Pushing Back Frontiers?. *Sound on Sound*. May. pp. 36–40.

Wiffen, P. 1995. E-mu Systems e64: Digital Sampler. *Sound on Sound*. July. pp. 26–27.

Wiffen, P. 2000. Second Sounds: E-mu Systems Emulator II 8-bit Sampling Keyboard. *Sound on Sound*. August. pp. 262–266.

Wiffen, P. and Scott, A. 1985. Me and My E-mu. *Electronics and Music Maker*. September. pp. 44–46.

Williams, E. 1982. The Rolls-Royce for Musicians. *Financial Times*. 4 August. p. 10.

Williams, N. 1982. Australian Synthesizer Cracks the World Market. *Electronics Australia*. August. pp. 30–34.

Williamson, J. & Cloonan, M. 2016. *Players' Work Time: A History of the British Musicians' Union, 1893–2013*. Manchester: Manchester University Press.

Wilson, S. 2009. Dubh it Again. *The Scotsman*. Wednesday 9 December. <www.scotsman.com/news/interview-the-skuobhie-dubh-orchestra-folk-band-1-771490> Accessed 24 January 2019.

Wyatt, S. 2003. Non-Users Also Matter: The Construction of Users and Non-Users of the Internet. in N. Oudshoorn and T. Pinch. eds. *How Users Matter: The Co-Construction of Users and Technology*. Cambridge: The MIT Press. pp. 67–79.

Yin, R. K. 2009. *Case Study Research: Design and Methods* (Fourth Edition). London: Sage.

Young, R. 2002. Worship the Glitch: Digital Music, Electronic Disturbance. in R. Young. ed. *Undercurrents: The Hidden Wiring of Modern Music*. London: Continuum.

Young, R. 2003. The Body Politician. *The Wire*. May. pp. 24–29.

Young, R. 2010. *Electric Eden: Unearthing Britain's Visionary Music*. London: Faber and Faber.

Zak, A. 2001. *The Poetics of Rock: Cutting Tracks, Making Records*. Berkeley: University of California Press.

Discography

ABC. 1982. *The Lexicon of Love*. Mercury.

AC/DC. 1980. *Back in Black*. EMI.

Akufen. 2001. *My Way*. Force Inc.

Art of Noise. 1983. Into Battle with the Art of Noise. ZTT.

Art of Noise. 1984. *(Who's Afraid of) The Art of Noise*. ZTT.

Eric B. & Rakim. 1987. *Paid in Full*. 4th & Broadway.

Afrika Bambaataa & the Soulsonic Force. 1982. Planet Rock. Tommy Boy.

Beastie Boys. 1986. *Licensed to Ill*. Def Jam.

Beastie Boys. 1989. *Paul's Boutique*. Capitol.

The Beatles. 1967. *Sgt. Pepper's Lonely Hearts Club Band*. EMI.

The Beatles. 1968. *The Beatles*. EMI.

Boogie Down Productions. 1987. *Criminal Minded*. B-boy Records.

James Brown. 1970. Funky Drummer (Part 1)/Funky Drummer (Part 2). King.

The Buggles. 1980. *The Age of Plastic*. Island.

Kate Bush. 1980. *Never For Ever*. EMI.

Kate Bush. 1982. *The Dreaming*. Fish People.

Daft Punk. 2001. *Discovery*. Virgin Records.

Daft Punk. 2013. *Random Access Memories*. Columbia Records.

Danger Mouse. 2004. *The Grey Album*. Bootleg.

De La Soul. 1989. *3 Feet High and Rising*. Tommy Boy.

Double Dee & Steinski. 1984. The Payoff Mix. Tommy Boy.

Dollar. 1982. *The Dollar Album*. WEA Records.

Dennis Edwards. 1984. Don't Look Any Further. Motown.

Todd Edwards. 1995. Saved My Life. FFRR.

Grandmaster Flash and the Furious Five. 1982. The Message. Sugarhill Records.

Found. 2006. *Found Can Move*. Surface Pressure Records.

Found. 2007. *This Mess we Keep Reshaping*. Fence Records.

Found. 2011. *factorycraft*. Chemikal Underground.

Frankie Goes to Hollywood. 1984. *Welcome to the Pleasuredome*. ZTT.

Peter Gabriel. 1980. *Peter Gabriel*. Charisma/EMI.

Paul Hardcastle. 1985. 19. Chrysalis.

Herbie Hancock. 1983. Rockit. Columbia.

Pierre Henry. 1966. *Variations for a Door and a Sigh*. Philips.

Herbert. 1998. *Around the House*. Soundslike/K7 Records.

Herbert. 2001. *Bodily Functions*. Soundslike/K7 Records.

Matthew Herbert. 2005. *Plat du Jour*. Accidental Records.

The Matthew Herbert Big Band. 2003. *Goodbye Swingtime*. Accidental Records.

The Matthew Herbert Big Band. 2008. *There's Me and There's You*. Accidental Records.

Matthew Herbert. 2010. *Mahler Symphony X*. Deutsche Grammophon Gesellschaft.

Matthew Herbert. 2011. *One Pig*. Accidental Records.

Matthew Herbert. 2013. *The End of Silence*. Accidental Records.

Jay-Z. 2003. *The Black Album*. Roc-A-Fella Records/Island Def Jam.

Chaka Khan. 1984. I Feel For You. Warner Brothers Records.

King Creosote. 2003. *Kenny and Beth's Musakal Boat Rides*. Domino Records.

King Creosote. 2004. *Red on Green*. Fence Records.

King Creosote. 2006. *KC Rules OK*. 679 Recordings.

King Creosote. 2007. *Bombshell*. 679 Recordings.

King Creosote. 2009. *Flick the Vs*. Domino Records.

King Creosote & Jon Hopkins. 2011. *Diamond Mine*. Domino Records.

Landscape. 1980. *From the Tea-rooms of Mars. . .*RCA.

Led Zeppelin. 1971. *Led Zeppelin*. Atlantic.

Malcolm McLaren. 1983. *Duck Rock*. Charisma/Virgin.

Matmos. 2001. *A Chance to Cut is a Chance to Cure*. Matador.

Nightcrawlers. 1992. Push the Feeling on (The Dub of Doom). Great Jones/ Island.

John Oswald. 1989. *Plunderphonic*. mLab.

John Oswald. 1993. *Plexure*. Avant.

Pink Floyd. 1973. *Dark Side of the Moon*. EMI.

Prefuse 73. 2003. *One Word Extinguisher*. Warp.

Public Enemy. 1991. *Apocalypse 91. . .The Enemy Strikes Back*. Def Jam.

Radio Boy. 2001. *The Mechanics of Destruction*. Accidental Records.

St Germain. 1995. Alabama Blues. F Communications.

DJ Shadow. 1996. *Entroducing*. Mo Wax.

Paul Simon. 1986. *Graceland*. Warner Brothers Records.

Sugarhill Gang. 1979. Rapper's Delight. Sugarhill Records.

Various Artists. 1976. *The Dartmouth Digital Synthesizer*. Folkways Records.

The Winstons. 1970. Color Him Father/Amen, Brother. Metromedia Records.

Wishmountain. 2012. *Tesco*. Accidental Records.

Stevie Wonder. 1979. *Stevie Wonder's Journey through the Secret Life of Plants*. Motown.

Stevie Wonder. 1980. *Hotter than July*. Motown.

Malcolm X. 1983. No Sell Out. Tommy Boy Records.

Yes. 1983. *90125*. Elektra.

Index

Printed in the United States
by Baker & Taylor Publisher Services